Hints To Travellers: Scientific And General, Volume 2

Edward Ayearst Reeves, Royal Geographical Society (Great Britain)

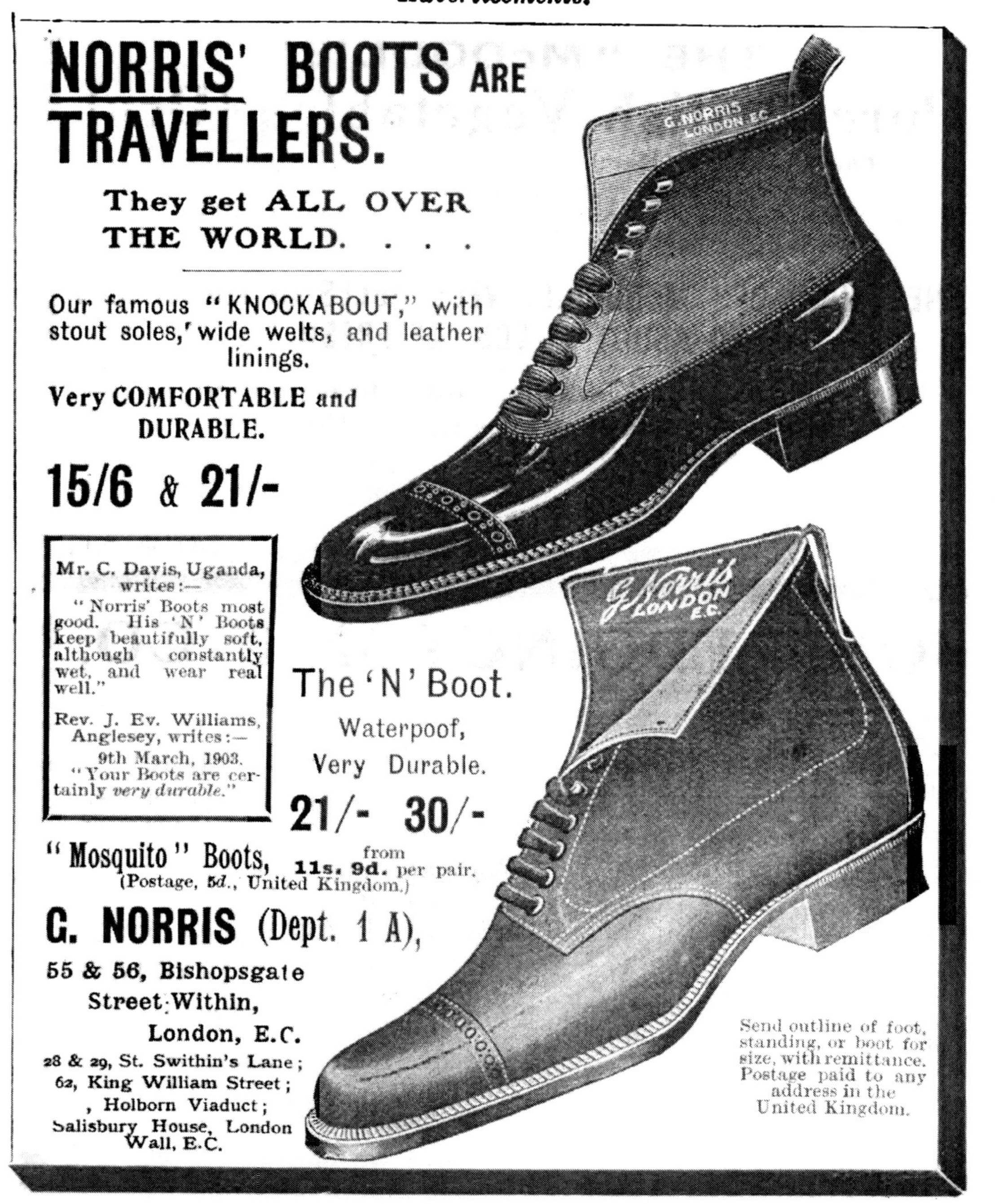
NORRIS' BOOTS ARE
TRAVELLERS.
They get ALL OVER THE WORLD. . . .
Our famous "KNOCKABOUT," with stout soles, wide welts, and leather linings.
Very COMFORTABLE and DURABLE.
15/6 & 21/-
G. NORRIS LONDON E.C.
Mr. C. Davis, Uganda, writes:—
"Norris' Boots most good. His 'N' Boots keep beautifully soft, although constantly wet, and wear real well."
Rev. J. Ev. Williams, Anglesey, writes:—
9th March, 1903.
"Your Boots are certainly very durable."
G. Norris LONDON E.C.
The 'N' Boot.
Waterpoof,
Very Durable.
21/- 30/-
"Mosquito" Boots, from 11s. 9d. per pair.
(Postage, 5d., United Kingdom.)
G. NORRIS (Dept. 1 A),
55 & 56, Bishopsgate Street Within,
London, E.C.
28 & 29, St. Swithin's Lane;
62, King William Street;
, Holborn Viaduct;
Salisbury House, London Wall, E.C.
Send outline of foot, standing, or boot for size, with remittance. Postage paid to any address in the United Kingdom.

HINTS TO TRAVELLERS

SCIENTIFIC AND GENERAL

EDITED FOR THE

Council of the Royal Geographical Society

BY

E. A. REEVES, F.R.A.S., F.R.G.S.

Instructor in Surveying and Practical Astronomy to the Royal Geographical Society.

NINTH EDITION

REVISED AND ENLARGED

VOL. II.

METEOROLOGY, PHOTOGRAPHY, GEOLOGY, NATURAL HISTORY, ANTHROPOLOGY, INDUSTRY AND COMMERCE, ARCHÆOLOGY, MEDICAL, ETC.

LONDON

THE ROYAL GEOGRAPHICAL SOCIETY

1, SAVILE ROW, W.

AND AT ALL BOOKSELLERS'

1906

Price of the two Volumes, 15s. *net*.

To Fellows, at the Office of the Society, 10s. *net*.

LONDON:
PRINTED BY WILLIAM CLOWES AND SONS, LIMITED,
DUKE STREET, STAMFORD STREET, S.E., AND GREAT WINDMILL STREET, W.

CONTENTS.

SECTION I.

SECTION II.

SECTION III.

SECTION IV.

SECTION V.

SECTION VI.

SECTION VII.

SECTION VIII.

MAPS AND ILLUSTRATIONS.

HINTS TO TRAVELLERS.

VOL. II.

I.

METEOROLOGY AND CLIMATOLOGY.

By HUGH ROBERT MILL, D.SC., LL.D., F.R.S.E.,
Secretary Royal Meteorological Society, Director of the British Rainfall Organization.

6

THE nature of the meteorological observations made by a traveller or by a resident in regions where there is no organised meteorological service will necessarily depend on the object which he has in view, the time he is able to devote to meteorological work, his knowledge of meteorology as a science, and his interest in it.

Of the many ways in which a traveller may add to the knowledge of atmospheric conditions, five may be specially mentioned :—

1. *A record of the weather, observed day by day with regard both to non-instrumental observations and the readings of instruments.* This may be taken as the minimum incumbent on all travellers.
2. *Observations for forecasting the weather and obtaining warning of storms.* This is sometimes of vital importance; it is always valuable at the time, and occasionally the results are worth recording. It may, however be looked upon as a practical application of the systematic observations.

3. *Observations with a view to determining the character of the local climate.* The traveller passing through a country can do little in this way, as long continued uniform observations in one place are necessary to fix the annual variations. Still, the recording of such data as may be obtained is always important in a little-known region, and the work of several travellers at different seasons will allow some fair deductions to be drawn. When a day is spent in camp, much importance attaches to regular observations made every two hours, from which the diurnal changes of climate may be ascertained.
4. *Special meteorological researches.* These usually demand special instruments and skilled observers. They naturally vary with the nature of the locality, *e.g.*, exact measures of radiation in deserts, of rainfall in forests and on adjacent open ground, of temperature during land and sea breezes, or of such conditions as fogs, thunderstorms, tornadoes, etc., in places subject to those visitations, are always of value. As a rule, however, the traveller cannot be expected to devote much time to these matters, unless the study of physical geography is the object of his journey.
5. *The collection of existing meteorological records.* It sometimes happens that at outlying stations meteorological observations have been taken and recorded for a considerable time. If they have not been already communicated to some meteorological society, the traveller should obtain a copy of them, and also compare the instruments in use with his own. He might in some cases aid in securing a knowledge of local climate by inducing residents at outlying stations to start regular observations.

The first two ways of advancing meteorology need alone be considered in detail; but with regard to all, it must be clearly understood that the value of the work is greater the more carefully the observations are made and recorded, and the more remote and less known the region.

1. A Record of Weather.—The traveller who makes his journey for any other purpose than the study of physical geography would be wise to burden himself as little as possible with instruments, but to understand thoroughly and use faithfully the few he carries. In a rapid march many different climates may be traversed in a few weeks, and the

records of variation of weather so obtained could not have much value; but when a halt of a few days or of a week or two is made, systematic observations become very valuable indeed.

Non-Instrumental Observations.

The first place must be given to non-instrumental observations, which may be made at any time on the march or in camp, and should always be noted at the time they are made in the rough note-book, and copied carefully into the journal each evening. These observations in the rough note-book will necessarily be mixed up with others on various subjects; but the meteorological facts should have a place reserved for themselves in the journal, say at the end of each day's work.

Wind.—Observations of the direction and force of the wind at several fixed hours in the day are advisable for comparison with instrumental readings; but on the march every decided change should be recorded if the nature of the country permits. In the depths of a forest, or in a narrow valley, the wind, if felt at all by the traveller, gives scarcely any clue to the movement of the air over the open country, but in most cases the movements of low clouds, when any are in sight, may be taken as a fairly satisfactory test. The direction is to be observed by means of the compass, and it will be sufficient to estimate it by the eight principal points—North, North-east, East, South-east, South, South-west, West, and North-west. Any sudden changes in direction so pronounced as to be noticeable should be recorded, for, taken in conjunction with the barometer readings, if the journey is along a route of nearly constant level, they are valuable aids in predicting the weather. In some places the direction of the wind has a well-marked regular diurnal change in perfectly settled weather.

Wind is always named by the direction from which it blows. The force of the wind is best estimated on the scale Calm, Light, Moderate, Fresh, Strong, and Gale. It is impossible, without long experience and the tuition of a trained observer, to assign relative numbers to these forces which should have any permanent value for comparison with the observations of others. Travelling on foot in a strong wind is always uncomfortable, and in a gale very difficult. If it is impossible to make way against the wind at all, or to pitch tents, the force may be put down as Hurricane after it has passed, the traveller bearing in mind that if he can write in his note-book at all, while unsheltered, a hurricane is not blowing. If a lake or a river without appreciable current is in sight, wind just sufficient to produce white crests on the waves may be called fresh, and that sufficient to blow away spray from the crests deserves to be termed strong. At sea, in a sailing-vessel, it is possible to acquire great skill in estimating wind-force; hence Beaufort's scale, originally devised with reference to the amount of sail a well-equipped frigate could carry, has come into extensive use, and it is as well to know it. By comparison with anemometers, the approximate velocity in miles per hour corresponding to the numbers on the scale has been estimated:—

BEAUFORT'S SCALE OF WIND FORCE.*

No.	Name.	Mean Velocity in miles per hour.
0	Calm	0
1	Light air	1
2	Light breeze	4
3	Gentle breeze	9
4	Moderate breeze	14
5	Fresh breeze	20
6	Strong breeze	26
7	Moderate gale	33
8	Fresh gale	42
9	Strong gale	51
10	Whole gale	62
11	Storm	75
12	Hurricane	92

The duration of strong wind should be noted, as well as the time of any marked change of strength. The land and sea winds of tropical coasts show a well-marked relation to the position of the sun and the hour of sunset, and in places where these winds blow the hours of calm and change should be noted. On mountain slopes a similar diurnal effect may be noticed; the wind usually blows uphill during the day and down-

* Equivalents obtained by Mr. R. H. Curtis using Köppen's method. See 'Symons's Meteorological Magazine,' 40 (1905), p. 157.

hill at night, while in valleys it usually blows either with or against the direction of the river. Local winds of peculiar character are sometimes met with in association with mountains such as the Föhn of the Alps, the Chinook wind of the Rocky Mountains, and the Helm wind of the Eden Valley in England and Adam's Peak in Ceylon.

Whirlwinds and tornadoes are rare phenomena, but if met with, it is worth while to take some trouble to put on record at least the hour of their appearance (local time), the direction in which the whirl moves onward, and the breadth of the path of destruction it leaves behind. When a storm of wind has passed over a wooded region and blown down many trees, the direction in which most of the trunks lie is worth observing. The top of the tree usually falls in the direction in which the wind was blowing, hence the root usually points to the direction of the wind. Waterspouts are closely allied to whirlwinds, and in any of those phenomena of revolving columns of air it is of much theoretical importance to determine the direction of the whirl about the axis, *i.e.*, whether the rotation is in the direction of the hands of a watch or the opposite. The prevailing wind of a district may often be discovered by the slope of trees growing on open ground, or still better, by the difference in the degree of wave erosion on small lakes. If the banks are of the same material all round, the side against which the prevailing wind drives the waves will always be the most worn away.

Cloud and Sunshine.—It would be impossible to keep a record of the countless changes in the cloud-covering of an English sky, but in many parts of the world the absence or presence of cloud is a function of latitude, altitude, and season, of great stability, and worthy of being attentively studied. The amount of cloud is usually estimated as the number of tenths of the sky covered; but it is a very difficult thing indeed to compare a tenth of the visible sky near the horizon with a tenth near the zenith. There is no difficulty, however, in observing when the sky is completely overcast or quite free of cloud, and as a matter of convenience the belt round the horizon to the height of thirty degrees may be neglected, *i.e.*, the lower third of the distance from the horizon to the zenith. Very often it will be found that clouds form and disappear at certain hours of the morning or evening, and it is useful to get exact information on the subject.

Of more importance than the amount of cloud is its nature, elevation, and movement. Distinct species of cloud have been recognised for a ong time, and from more recent studies it would appear that they owe their distinctive appearance to the altitude at which they float in the air. Meteorologists distinguish a number of classes and transitional forms of cloud; it is enough for the traveller to be able to recognise the most definite types, viz., Cirrus, Cumulus, Stratus, and Nimbus. *Cirrus* clouds are the small tufts or wisps of cloud which float very high in the atmosphere, and to which the popular name of "mare's tails" is applied. The transitional form, *Cirro-Cumulus*, popularly known as "mackerel scales" or "mackerel sky," is equally easy to identify. *Cumulus* clouds are great woolly-looking heaps of cloud, the lower surface of which is often nearly horizontal, while above they well into an exuberant variety of rounded forms. They represent the condensation of moisture in ascending columns of heated air. *Stratus* clouds are low-lying sheets of condensed moisture, which, being usually seen at a low angle, appear like thin layers parallel to the horizon. The transitional type *Cirro-Stratus* is usually seen in the form of great feather-like clouds stretching across nearly the whole sky. *Nimbus* is a rather low-lying cloud from which rain is falling even if the rain is re-evaporated before reaching the ground. The lowest clouds of all, those resting on the surface of the ground and enveloping the observer, are called *mist* and *fog*. The two are distinguished by the fact that a mist wets objects immersed in it, while a fog does not. All varieties of cloud are physically the same, consisting of minute globules of liquid water falling through a portion of air saturated with moisture. The globules being small offer a relatively great surface to friction, and so fall very slowly, and in the higher clouds they evaporate on the lower surface before they have time to agglomerate into raindrops. In the highest of all clouds, the cirrus type, the particles are spicules of ice and not globules of water. It is a common error to suppose that black clouds differ from white clouds. All clouds are white when they reflect the light of the sun, and all are black when they come between the eye and the sun in sufficient thickness to cut off a considerable portion of its light.

The sudden appearance of a particular kind of cloud is important as a weather sign. It shows that changes are going on in the vertical circulation of the atmosphere. Hence if cirrus or cumulus cloud should be observed to be increasing the fact should be noted, and the direction in which the clouds are moving should be noted also.

In observing cloud-motion attention should be given only to the sky overhead; at any lower angle the parallax due to viewing the clouds obliquely deprives the observation of value. It is also necessary to distinguish between the movement of the upper and of the lower clouds, as these are floating in very different parts of the atmosphere. It is comparatively rarely that the motion, say of nimbus and cirrus, is in the same direction. On a lofty mountain, strata of cloud which from below were seen to be cumulus may be passed through as layers of mist, and on emerging from them their upper surface may be seen below one. In many mountains the cloud-belt is as sharply defined as the snow-line, and its variations should be carefully observed.

Clouds should occasionally be photographed as a record. This should be done especially when a type of cloud comes to be recognised as a usual one, for while exceptional forms may prove interesting, a record of the usual forms is certain to be valuable. In this connection a protest may be made against the horrible custom of some amateur and of many professional photographers of printing in clouds from some stock negative in their pictures of scenery. The cloud is an essential part of a picture, and it is better to leave an over-exposed sky of natural cloud than to insert a beautiful representation of a cloud-form which may be one never visible in the particular place or at the particular season.

Mist, Fog and Haze.—Mist or fog at low levels will of course be recorded whenever observed, and its density and duration noted. A good way to define the density of thick fog is to measure the number of yards at which an object becomes indistinguishable, and the most convenient object for the purpose is a person. Light mists lie over water or marshes at certain hours in particular seasons, and their behaviour should always be observed. It often happens that the distant view from a height is obscured by a haze not due to moisture, and this appearance should be noticed with a view to discovering its cause. The smoke from forest or prairie fires in Canada sometimes produces so thick a haze as to put a stop to surveying operations for weeks at a time. Haze is often due to dust blown from deserts, or ejected from volcanoes, and sometimes to swarms of insects.

Rain and Dew.—The journals of most travellers fail to give a clear idea of the prevalence of rain during their journeys, and it is much to be desired that something more explicit than "a showery day" or "fairly dry" should be recorded. The hour of commencement and cessation of rain during a march should be noted, and some indication given as to whether the rain fell heavily or lightly. In this way any tendency to a diurnal periodicity of rain would be detected, and some definite meaning would be given to the terms rainy season and dry season. If rain occurs during the night it should also be recorded, and the amount of night rains should always be measured by means of a rain gauge in the manner to be described later.

The general condition of a country with regard to rain may often be judged from the appearance of vegetation or the marks of former levels of high-water in lakes or rivers. Thus on mountain slopes or the sides of a valley any difference in the luxuriance of vegetation according to exposure probably indicates the influence of rainfall as guided by the prevailing wind. So, too, the appearance of lines of drifted débris on the banks some distance from the edge of a lake or river may be taken as indications of the height to which the waters sometimes rise; and conversely the appearance of rows of trees in the middle of a wide shallow lake may indicate the line of a river which has temporarily flooded the surrounding meadows. Such observations have an important bearing on climate.

The appearance and amount of dew are also to be recorded. The most important points to notice are the hour in the evening when the deposit commences, and the hour in the morning when the dew disappears. It should be noted also whether the deposit of dew is in the form of small globules standing apart on exposed surfaces, or if it is heavy enough to run together into drops and drip from vegetation to the ground.

Thunderstorms and Hail.—The occurrence of thunderstorms should of course be noted, and here the hour of occurrence is of very great importance, for thunderstorms frequently show a marked diurnal period. The appearance of lightning without thunder should be recorded when it is observed, but this will naturally be almost always after sunset. Hailstorms usually accompany thunderstorms, and sometimes take the place of them. The occurrence of hail is most frequent in summer, and records of the size of hailstones are important. If possible they should, when very large, be photographed along with some object of known size, and their structure described. It might at least be noticed whether they are hard and clear, like pure ice, or opaque like compacted snow, or made up of concentric layers of clear and opaque ice alternately.

Snow.—Snow falls in all parts of the world, although in tropical or subtropical latitudes only at great elevations above sea-level. The actual limits of snowfall at sea-level are as yet imperfectly known, and any observations of snow showers in the neighbourhood of the tropics are of importance. It is essential in such a case to record also the approximate elevation of the land. On mountains in all latitudes the position of the snow-line should be noted at every opportunity. This is the line above which snow lies permanently all the year round, or below which snow does not completely melt in summer; and it is a climatic factor of some importance. It may be remarked, for instance, that if the traveller finds snow lying on grass, moss, or other vegetation, he is certainly not above the snow-line. It is necessary also to notice that glaciers may descend unmelted a long distance below the level of perpetual snow. While the conditions of snow lying on the ground in the Arctic regions and above the snow-line in any part of the world are matters pertaining more to geology and mountaineering than to meteorology, the duration of snow-showers, the character of the snow, and the depth to which it lies on ground below the snow-line are too important from their bearing on climatology to be overlooked.

The character of the snow as it falls varies from the sleety, half-melted drops common in warm air to the fine dust of hard, separate ice-crystals found in the intense cold of a Polar or Continental winter. The feathery appearance of lightly-felted flakes is an intermediate type between the two extremes. In measuring the depth of snow as it lies, care should be taken to select open ground where there is no drifting, and when the snow is not too deep the measurement can usually be best made with a walking-stick on which a scale of feet and inches (or of centimetres) has been cut. Such a stick is useful for measuring the depth of shallow streams, and for many other purposes. The result should be entered as "depth of fallen snow," so that there may be no risk of confusing the figures with the amount of snowfall estimated as rain. Speaking roughly, a foot of snow is usually held to represent an inch of rain. A violent storm of wind, accompanied with falling snow, is termed a *blizzard* in the western United States, and a *buran* in Siberia; but in recording such a phenomenon it is better to describe its nature than to give it a name which may possibly be misleading.

Frost.—The appearance of frost in the form of hoar-frost (the way in which atmospheric water-vapour is deposited in air below the freezing-point), or of thin ice formed on exposed water, should always be carefully looked for and noted. In hot, dry countries the intense radiation from the ground at night often reduces the temperature below the freezing-point, although, during the day, the ground may be very hot. The appearance of frost at sunrise is a valuable check on the readings of a minimum thermometer, and in most cases is a more trustworthy datum. Similarly in cold countries, where snow is lying on the ground or ice covering the rivers, the appearance of thaw, especially in cloudy weather, is a delicate test of the rise of the air temperature to the freezing-point. The traveller should never fail to record cases of melting and solidifying of any substances due to changes of temperature. The softening of candles and the freezing of mercury or of spirits give information regarding temperature at least as valuable as the readings of thermometers.

Other Observations.—Any peculiar atmospheric phenomena, such as the appearance of the zodiacal light after sunset, the aurora, the electrical lights seen on pointed objects, and known as St. Elmo's fire, rainbows, especially lunar rainbows, haloes, the appearance of mock-suns or moons, meteors or shooting stars, should be noted on their occurrence, as many of them are valuable weather prognostics. Attention should also be given to any appearances of mirage, or other effects of irregular distribution of atmospheric density. A mirage is only rarely so perfect as to show ships inverted in the air, palm-grown islands in the sea, or distant oases in the desert. The common form is an unusual intensification of refraction, raising land below the horizon into sight, or apparently cutting off the edges of headlands or islands at sea or on large lakes. It is worth while observing the temperature of the air and of the water or ground when an unusually clear mirage effect is visible.

Another interesting series of observations may be made on the colours of the sky and clouds at sunrise and sunset. A phenomenon often observed at sunset, but the existence of which is still sometimes denied, should be looked for. This is the appearance of a gleam of coloured light at the moment when the upper edge of the sun dips below a cloudless horizon. A note should be made of the nature of the horizon, whether land or sea, and of the colour of the light if it should be observed. When opportunity offers, the first ray of the rising sun might be similarly observed.

The traveller should, at the end of each day, give his opinion of the nature of the weather, saying whether he felt it hot or cold, relaxing or bracing, close or fresh. Such observations have no necessary relation to degree of temperature or humidity as recorded by instruments; but the human body is the most important of all instruments, and everything which affects it should be studied. By paying attention to the foregoing instructions, an observant traveller will bring home a far better meteorological log without instruments than a more careless person could produce by the diligent reading of many scales. Yet, in enforcing the importance of non-instrumental observations, we must not leave the impression that the readings of instruments are of little value. It is, in ordinary circumstances, only by the readings of instruments that the climate of one place can be compared with that of another, and only the best results of instrumental work are precise enough to form a basis for climatological maps.

Instrumental Observations.

The minimum requirement of instrumental observations by a traveller is the reading twice daily of the barometer and of the dry and wet bulb thermometers, to ascertain the temperature and humidity of the air, also the reading once daily, in the morning, of the minimum thermometer which has been exposed all night, and on days in camp of the maximum thermometer also. It is very desirable to expose a rain gauge whenever it is practicable to do so. Unless special meteorological researches are to be carried out, nothing farther in the way of observations need be attempted. A very useful supplement to the necessary observations is the use of a self-recording barograph or thermograph; but these are delicate instruments, liable to get out of order unless very carefully handled, and it will not always be possible to make use of them.

The observer must understand what his instruments are intended to measure, how they act, and how they should be exposed, read, and the reading recorded. He must know enough about all these things to be able to dispense with unnecessary precautions only possible at fixed observatories, and, at the same time, to neglect nothing that is necessary to secure accuracy in the results.

Thermometer Corrections.—All thermometers, without exception, should have the degree marks engraved on the stem, or on a slip of enamel within the outer tube, and be supplied with a certificate from the Kew Observatory showing what the error of the scale is at different points. This certificate should be in duplicate, and a copy ought to be left in a safe place at home. After a long journey the thermometers which have been in use should be sent to have their errors re-determined. The corrections are not, however, to be applied by the observer unless he is working out his observations for some special purpose. No thermometer is passed at the Kew Observatory if its error approaches one degree in amount, so that for all ordinary purposes of description a Kew certificated thermometer may be looked on as correct. But when the readings are being critically discussed and compared with the observations of other people, the correction is of the greatest importance. *It cannot be too strongly impressed upon an observer that, in reading meteorological instruments, he must read exactly what they mark, and record that figure in his observation-book on the spot.* The corrections can be applied afterwards by the specialist who discusses the work. For subsequent reference it is necessary to note in the observation-book the registered Kew number of the thermometer in use, and if a thermometer should get broken and another be used instead, the number of the new instrument must be noted at the date where it is first employed. Care should be taken to use the same thermometer for one purpose all the time if possible, and only an accident to the instrument should necessitate a change being made.

Thermometers are either direct-reading or self-registering. The former are used for obtaining the temperature at any given moment, the latter for ascertaining the highest or the lowest temperature in a certain interval of time. They are filled either with mercury, or a light fluid which freezes less readily, such as alcohol or creosote.

Thermometer Scales.—The particular system on which the thermometers are graduated is of no importance, but merely a matter of convenience. The Fahrenheit scale is used for meteorological purposes in English-speaking countries; but for all other scientific purposes the Centigrade scale is used everywhere. One can be translated into the other very simply by calculation*; but it is convenient for a traveller to have all his thermometers graduated in accordance with one scale only.

The graduation, as marked on the stem of the thermometer, is usually to single degrees, but anyone can learn to read to tenths of a degree by a little practice. Care must be taken to have the eye opposite the top of the mercury column. Suppose it to be between 50 and 51, the exact number of tenths above 50 is to be estimated thus: If the mercury is just visible above the degree mark it is 50°·1, if distinctly above the mark 50°·2, if nearly one-third of the way to the next mark 50°·3, if almost half-way 50°·4, exactly half-way 50°·5, a little more than half-way 50° 6, about two-thirds of the way 50°·7, if nearly up to the next mark 50°·8, and if just lower than the mark of 51° it is 50°·9. The eye soon becomes accustomed to estimating these distances.

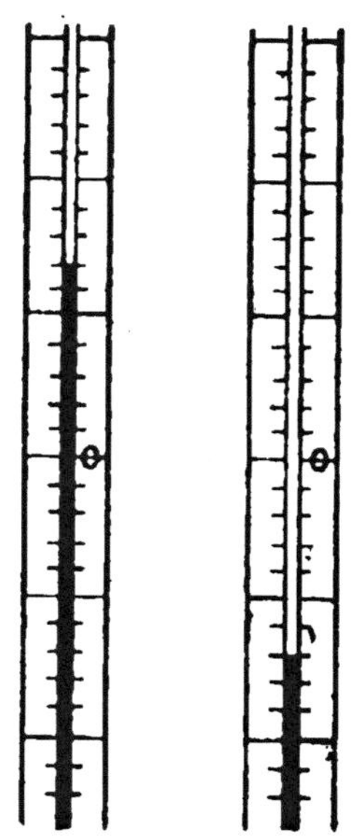

Fig. 1.—*Reading Thermometer Scale above and below Zero.*

In using a thermometer below zero, the observer must pay attention to the change in the direction of reading the scale, the fractions of a degree counting downward from the degree mark instead of upward from it, as in readings above zero. Readings below the zero of the scale are distinguished in recording them by prefixing the minus sign. The annexed figure shows the reading of two thermometers graduated to fifths of a degree, one showing a temperature of 1°·4, the other of −1°·4. The importance of remembering that the scale is inverted at the zero point is of course greater in the case of a Centigrade thermometer, where the zero is at the freezing point of water.

Care of Thermometers.—Mercurial thermometers will always be employed for ordinary purposes in places where the temperature is not likely to fall to −40°: *i.e.*, everywhere except in

---

* For comparison of scales see Table XXVI., Vol. I., p. 333.

To convert Centigrade readings into Fahrenheit the rule is "Multiply by 1·8 and add 32." This can be done mentally in a moment, thus: "Multiply by 2, subtract one-tenth of the product, and add 32." *E.g.*, to convert 10° C. to F.: $10 \times 2 = 20$; one-tenth of $20 = 2$; $20 - 2 = 18$; $18 + 32 = 50°$ F.

the polar regions and the interior of continents north of 50° N. These thermometers are very strong and are not easily broken except by violence. The one vulnerable part is the bulb, which is of thin glass and filled with heavy mercury. Hence, in carrying thermometers, care has to be taken to protect the bulb from coming in contact with any hard object. The best way to carry an unmounted thermometer is in a closed brass or vulcanite tube with a screw top, the inside of the tube being lined with india-rubber and provided with a cushion of cotton-wool for the bulb to rest on. If the thermometer is mounted in a wooden frame it should be secured in a box so that the frame is firmly held and the bulb projects into a vacant part of the box, which may be lightly filled with cotton-wool or provided with a deep and well-padded recess. Every thermometer which is not graduated above 120° should have an expansion at the top of the tube which the mercury that may be driven beyond the scale by over-heating will not fill; otherwise any accidental over-heating will break the bulb.

The unavoidable shaking or any sudden shock during travelling is apt to cause the mercury column to separate, and a portion of it may be driven to the top of the tube, where it may remain unless looked for and brought back. Hence it is important to see that the top of the bore of the tube is visible, and not covered by any attachment holding the tube to a wooden frame. Thermometer readings are absolutely valueless unless the whole of the mercury fills the bulb and forms a continuous column in the stem. To bring a broken column together the best plan is to invert the thermometer, if necessary shaking it gently, until the mercury flows from the bulb and entirely fills the tube, leaving a little vacant dimple in the mass of mercury in the bulb. When this is done, the thermometer should be brought into its normal position bulb downwards, and the column will usually be found to have united. If this method does not succeed the thermometer may be held in the hand by the upper end, raised to the full stretch of the arm, and swung downwards through a wide arc with a steady sweep. I have never known this method to fail.

Thermometer Screens.—It is usual at fixed stations to expose the thermometer to the air by hanging it in a screen made of louvre-boards so arranged that the air penetrates it freely while the direct rays of the sun are cut off. The Stevenson screen, constructed on this plan with a door opening on the side away from the sun, is well adapted for use in temperate countries; but it is too cumbrous to carry on a journey and does not afford sufficient ventilation for use in tropical countries. An excellent substitute is the canvas screen devised by the late Mr. H. F. Blanford, which consists of a bamboo frame carrying the thermometers (with their bulbs four feet from the ground). The whole structure is five feet high, and is sufficient for any places where the wind is moderate. It is constructed of bamboos or rods of light wood, cords, and canvas, which may readily be made up before starting, and it is easily renewed

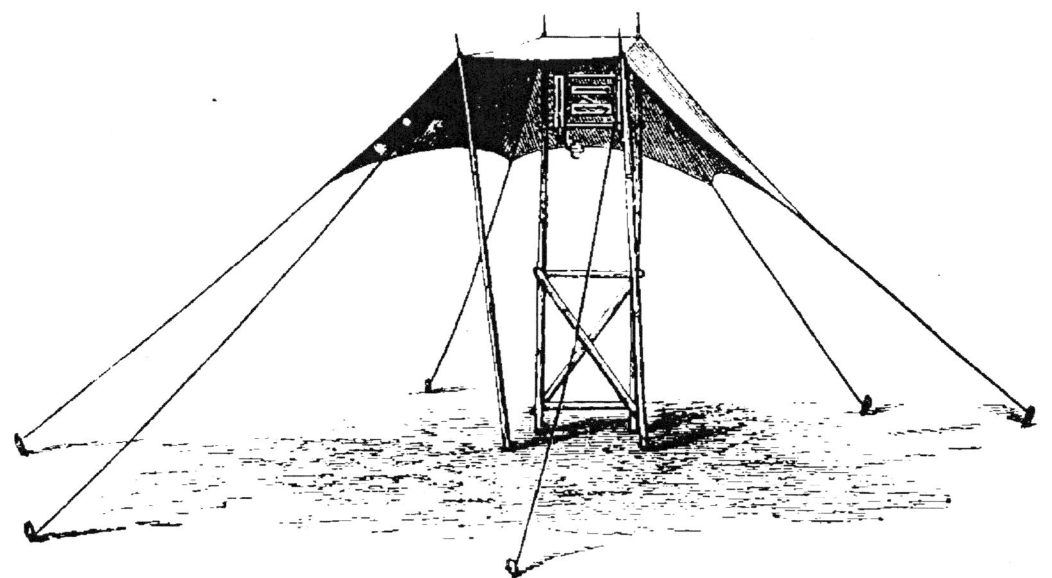

FIG. 2.—*Mr. H. F. Blanford's Portable Thermometer Screen.*

or repaired. The canvas roof should be triple or quadruple according to the thickness of the material. Such a screen will afford sufficient protection at night, or even in the day, if set up in the shade, and it will throw off rain; but in the sun it will require a thick mat as an additional protection on or preferably stretched above the roof.

For a more permanent station the form of exposure recommended by a committee of the British Association for use in tropical Africa will be found very suitable in hot countries.

The thermometers are placed in a galvanised iron cage, which is kept locked for safety. This cage is suspended under a thatched shelter, which should be situated in an open spot at some distance from buildings. It must be well ventilated, and protect the instruments from being exposed to sunshine or rain, or to radiation from the ground. A simple hut, made of materials available on the spot, would answer this purpose. Such a hut is shown in the drawing (Fig. 2). A gabled roof with broad eaves, the ridge of which runs from north to south, is fixed

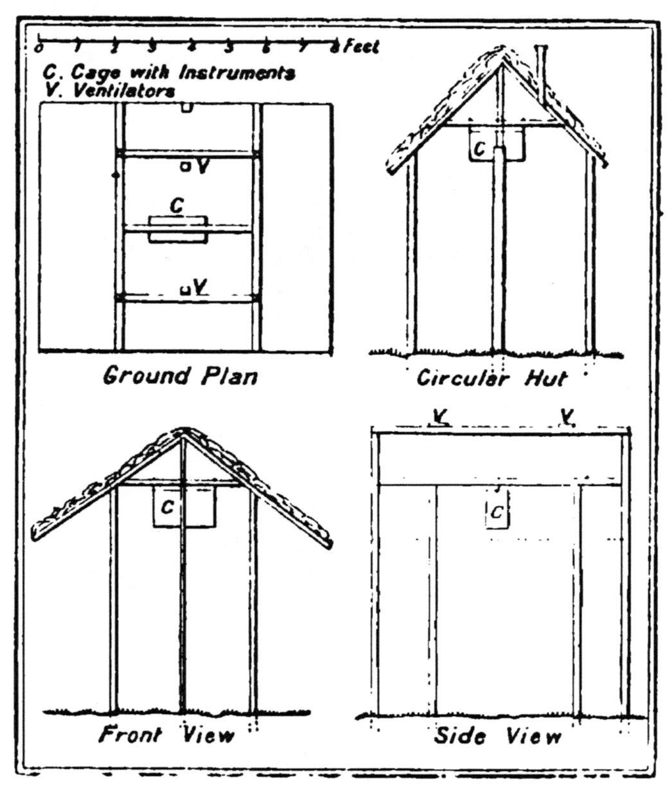

FIG. 3.—*Hut for Sheltering Thermometers.*

upon four posts, standing four feet apart. Two additional posts may be introduced to support the ends of the ridge beam. The roof at each end projects about eighteen inches; in it are two ventilating holes. The tops of the posts are connected by bars or rails, and on a cross bar is suspended the cage with the instruments. These will then be at a height of six feet above the ground. The gable-ends may be permanently covered in with mats or louvre-work, not interfering with the free circulation of the air, or the hut may be circular. The roof may be covered with palm-fronds, grass, or any other material locally used by the natives for building. The floor should not be bare but covered with grass or low shrubs.

The great object of these precautions is to obtain the true temperature of the air, and avoid the excessive heating due to the direct rays or reflected heat of the sun falling on the thermometers, and the excessive cooling due to the radiation of heat from the thermometers to a clear sky at night. Such a shelter is absolutely necessary when maximum and minimum thermometers are used; but can be dispensed with for the simple observation of the temperature of the air at a given time. This may be effected by securing a rapid flow of air over the thermometer, either by causing the air to flow past the instrument or by causing the instrument to move rapidly through the air. It has been found by experiment that the true temperature of the air is obtainable in this way whether the operation is performed in sunshine or in shade; but it is preferable to do so in the shade.

Sling Thermometer.—The sling thermometer is the most simple and convenient of all instruments for ascertaining the temperature of the air. It is an unmounted thermometer with a cylindrical bulb, and the degree-marks engraved on the glass stem. The upper end terminates in a ring to which a silk cord about two feet long is attached. As a precaution it is as well to secure the cord by a couple of clove hitches round the top of the thermometer stem as well as to the ring, as the thermometer would then be held securely even if the ring broke. The thermometer is used by whirling it in a vertical circle about a dozen times, the observer taking care, by having a loop of the string round the wrist or finger, that it is not allowed to fly off. Then the thermometer is read, swung once more, and read again. This process is repeated until two consecutive readings are identical; when this is the case the instrument shows the true temperature of the air. It is sufficient to note the final temperature in the observing book.

The risk of breaking a sling thermometer is the only drawback to its use. Only a silk cord should be used, and it should be examined frequently to see that it has not got chafed. In swinging the thermometer, an open place must be selected where it is not likely to come in contact with a branch or any other object.

Hygrometers.—As the humidity or degree of moisture in the atmosphere is a very important climatic factor it is necessary to measure it as carefully and as frequently as the temperature of the air. There are many instruments, called psychrometers or hygrometers, for doing this; but few of them are simple enough for the use of a traveller. The proportion of water-vapour in air is a little difficult to understand at first, because it is not a constant quantity as in the case of the other constituents of air, but varies according to the amount of water-surface exposed to the air and according to the temperature. The maximum amount of water-vapour which can be present in air varies with the temperature, being greater as the temperature is higher and less as the temperature is lower. Thus, if air at 50° F. contains the maximum amount of water-vapour which it can contain at that temperature, it is said to be *saturated*, for it will take up no more and evaporation stops; and if the temperature were to fall ever so little there would be more water-vapour present in the air than it could hold and some would separate out and condense into dew or rain, hence the temperature of saturation is called the dew-point. But if air saturated at 50° is warmed up say to 60° it can then contain more water-vapour than it has, and the temperature would require to fall 10° before dew or rain could form. When the air is not saturated water exposed to it evaporates rapidly until the maximum quantity of water-vapour is again present, a larger quantity corresponding to the higher temperature. At any given temperature the essential thing to know about the humidity of the air is the additional amount of water-vapour it could take up before becoming saturated, or in other words the humidity relative to the maximum humidity possible at the existing temperature. The relative humidity is expressed in percentages of the maximum humidity possible (saturation) at the actual temperature of observation. It may be measured by two methods, (1) finding the dew-point or temperature at which the amount of vapour present saturates the air; (2) by finding the rate at which the air allows evaporation to proceed; the farther the air is from saturation the more rapid is the rate.

The dew-point may be found directly by means of an instrument by which the air is cooled down until it begins to deposit moisture on a polished surface, but such an instrument is inconvenient to handle when travelling. It may also be found indirectly by calculation from the relative humidity.

The relative humidity is most easily calculated from the rate of evaporation. It is one of the laws of evaporation that heat is required to change liquid into a vapour, and when evaporation is going on heat is being abstracted from surrounding bodies, and they are growing colder. By allowing evaporation to take place from the bulb of a thermometer the rate of evaporation may be measured by the fall of temperature produced, and tables have been constructed to convert the differences between the wet and dry bulb readings into relative humidities.

The wet-bulb thermometer consists of an ordinary thermometer, the bulb of which is covered with clean muslin and kept moist by means of a piece of cotton lamp-wick dipping into a small vessel of pure water. Care must be taken to have the water quite pure and free from salt, otherwise the true reduction of temperature will not be observed. Hence special precautions are necessary when observing at sea or in an arid country where the ground is covered with incrustations of salt.

In any form of wet bulb thermometer when the air is much below the freezing point, it will usually be found most satisfactory to remove the muslin covering and allow the bulb to become covered with a coating of ice, by dipping it into water and allowing the water to freeze upon it. Evaporation takes place from solid ice sufficiently rapidly to give the true wet-bulb readings at least with a sling thermometer.

When the air is saturated, *i.e.*, relative humidity = 100 per cent., there is no difference in the reading of the wet and dry bulb thermometers, and the greater the difference between the readings the smaller is the relative humidity of the air.

The wet-bulb thermometer has to be exposed to the air with the same precautions as are taken in the case of the dry bulb. The two may be hung side by side—but at least six inches apart—in the screen or cage described on p. 15; or the wet bulb may be employed as a sling thermometer. One way to do this is to tie a muslin cap on the bulb of the sling thermometer with a piece of wet lamp-wick coiled round the upper part of the bulb, and then whirl it until the reading becomes constant, taking care to moisten the bulb again if it should become dry. Another way is simply to twist a piece of filter-paper or blotting-paper round the bulb, and dip it in water before swinging.

Aspiration Psychrometer.—Perhaps the most convenient form of wet and dry bulb thermometer for use by a traveller is that known as

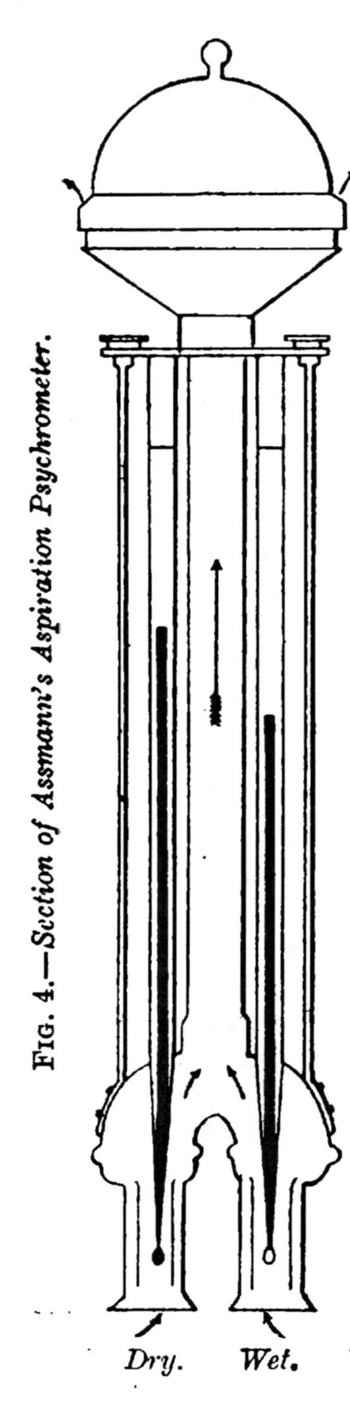

Fig. 4.—*Section of Assmann's Aspiration Psychrometer.*

Assmann's Aspiration Psychrometer. It requires no protecting screen, is not subject to the risk attending the use of the sling thermometer, and gives an extremely close approximation to the true temperature and humidity. The principle of the instrument is very simple. The wet and dry bulb thermometers are enclosed separately each in an open tube (see Fig. 4) through which a current of air is drawn by means of a fan, actuated by clockwork in the upper part of the case. In making an observation, all that is required is to see that the water vessel for the wet bulb is filled and the bulb properly moist, and that the dry bulb is free from any condensed moisture. The instrument is then hung to a branch or other support placed in the open air (or even held in the hand), preferably in the shade, although this is not essential, and the clockwork wound up. Air will then be drawn over the bulbs for five minutes or more, and if the temperature of each thermometer has not become steady by the time the clockwork has run down, it must be wound up again.

The thermometers in Assmann's Psychrometer are graduated according to the Centigrade scale, and each degree is subdivided into fifths on a slip of porcelain enclosed in the outer tube of the thermometer (see p. 18).

Minimum Thermometer.—The minimum temperature of the night can usually be ascertained by a traveller exposing a minimum thermometer when the camp is set up and reading it in the morning before starting on his way. There are several forms of minimum thermometer, but the only one likely to be used is that known as Rutherford's. It is very delicate and liable to go out of order. The instrument should be of full size, as used in meteorological stations at home; it must be packed so as to be as free as possible from shock or vibration, and ought to be carried in a horizontal position. The bulb is filled with alcohol or some similar clear fluid, and within the column of spirit in the stem there is included a little piece of dark glass shaped like a double-headed pin. This is the index which continues pointing to the lowest temperature until the instrument is disturbed or re-set. The thermometer has to be hung in a horizontal position. When the temperature rises, the column of spirit moves along the tube, flowing past the index without disturbing it. When the temperature falls, the spirit returns towards the bulb, flowing past the index *until the end of the column touches the end of the index.* The phenomenon known as surface-tension gives to the free surface of any liquid the properties of a tough film, and the smaller the area of a free surface is, the greater is this effect of surface-tension. Hence it is that the inner surface of the column of alcohol is not penetrated by the glass index, but draws the index with it backwards towards the bulb. As soon as the temperature begins to rise, the alcohol once more flows past the index towards the farther end of the tube. The end of the index farthest from the bulb remains opposite the mark on the stem indicating the lowest temperature which had occurred since it was last set, and this reading must be taken without touching the thermometer.

To set the index it is only necessary to tilt the bulb end of the tube upwards, when the index will slide down by its own weight until it comes in contact with the inner surface of the end of the column of alcohol.

Care of a Minimum Thermometer.—The chief dangers to which a minimum thermometer are liable are three--(1) the index being shaken into the bulb, (2) the index being shaken partly or wholly out of the column of spirit, and sticking in the tube, and (3) the column of spirit becoming separated or a portion of the spirit evaporating into the upper end of the tube.

The thermometer should be so constructed as to make it impossible for the index to get into the bulb, or with an index so long as not wholly to leave the tube, and this should be seen to before purchasing. When any of the other derangements occurs the natural instinct of an observer is to immerse the thermometer in warm water until the spirit entirely fills the tube, and then allow it to cool. The only drawback to this simple method is the almost inevitable bursting of the bulb and destruction of the thermometer. This method should never be attempted; but if the warning were not given, the idea would be sure to occur to the observer some time or other, and he would proceed to destroy his thermometer with all the fervour of a discoverer. The only satisfactory way to rectify a deranged minimum thermometer is as follows:

If the column is separated, but the index remains in the spirit, grasp the instrument firmly by the upper end and swing it downwards with a jerk (as in the case of the mercurial thermometer mentioned on p. 23). If the index has been shaken out of the spirit and remains sticking in the upper part of the tube, or if a little spirit has volatilised into the top of the tube and cannot be shaken down by the first method, a quantity of spirit should be passed into the upper end of the tube by grasping the thermometer by the bulb end of the frame and swinging in the same way. When the index is immersed or the drop of volatilised spirit joined on to the column, the first process of swinging by grasping the upper end of the tube will bring the instrument into working order. After any operation of this kind the thermometer should be kept in a vertical position bulb downwards, to allow the spirit adhering to the sides of the tube to drain back completely. Then the thermometer should be brought into the horizontal position and set by allowing the index to slide down to the end of the column of spirit. The end of the column of spirit farther from the bulb should always show the same temperature as the dry-bulb thermometer. If it should be observed to read a degree or two lower, it will be found that some of the spirit has volatilized and condensed at the end of the tube.

The minimum thermometer should be exposed to the air four or six feet from the ground under a screen or roof, like that described on p. 15, so that it is not exposed to the open sky, and the ground under the shelter should be covered with grass or leaves, not on any account left bare. The loss of heat by radiation of the ground to the open sky will produce a night temperature much lower than that of the air a few feet above the ground, and a radiation thermometer is often employed, laid on the grass and exposed to the sky to measure this effect.

Travellers, however, can rarely be expected to make observations of such a kind, as the instrument is one of extreme delicacy.

Maximum Thermometers.—Maximum registering thermometers are filled with mercury, and are less liable to get out of order than spirit-thermometers. The simplest and best form for use by travellers is Negretti and Zambra's. Its principle is very simple. When the temperature rises and the mercury in the bulb expands, it forces its way along the stem in the usual manner; but there is a little constriction in the tube just outside the bulb which breaks the column as the temperature begins to fall, and so prevents the mercury in the bulb from drawing back the thread of mercury from the tube. The thermometer is hung horizontally, and the end of the mercury farthest from the bulb always shows the highest temperature since it was last set. Before reading the thermometer, it is well to take the precaution of seeing that the inner end of the thread of mercury is in contact with the constriction in the tube, and if, by the shaking of the instrument or otherwise, the mercury has slipped away from this position, it should be brought back to it by tilting the thermometer bulb downwards very gently, then returning it to the horizontal position and reading.

To set this thermometer, it is only necessary to hold it vertically bulb downwards and shake it slightly, if necessary striking the lower end of the frame carrying the instrument, gently against the palm of the hand. This causes the mercury to pass the constriction and re-enter the bulb. When set, the end of the column farther from the bulb should indicate the same temperature as the ordinary dry-bulb thermometer.

Another form of maximum thermometer is known as Phillips'. It is an ordinary mercurial thermometer, but a short length of the upper part of the column in the tube is separated from the rest by a little bubble of air. It is used in the horizontal position, and as the temperature rises the whole column moves forward, while, when the temperature falls, only that portion behind the air-bubble retires towards the bulb. The tip of the column thus remains to mark the maximum temperature to which its farther end points. The instrument is set by gently tilting the bulb end downwards, when the detached portion of the column at once runs back until stopped by the air-bubble. This is the most convenient instrument to use at a fixed station; but in travelling it is apt to get out of order as shaking may have the effect of allowing the air-bubble to escape into the upper part of the tube, or into the bulb, and the instrument cannot easily be brought into working order again.

Rain-Gauge.—While measurements of rainfall can possess no climatological value unless they are carried on continuously at a fixed station, some very interesting observations may be made by the traveller both during the night when in camp, and during heavy showers when compelled to stop on the march. The rain-gauge is in itself the most simple of all scientific instruments, for it consists merely of a copper funnel to collect the rain as it falls, and a bottle to contain what has been collected. A graduated measuring glass is the only accessory required. Rain is measured by the depth to which the water would lie on level ground if none soaked in, evaporated or flowed away. On an emergency, a rain-gauge can be improvised out of a biscuit tin, or any vessel with vertical sides and an unobstructed mouth. Such a vessel standing level would collect the rain, the depth of which might be measured by an ordinary inch-rule. It is rare, however, to find rain so heavy as to give any appreciable depth when collected in a vessel freely open to evaporation, and in order to estimate the amount of rainfall to small fractions of an inch, the device is employed of measuring the water collected in the receiver of the gauge in a glass jar of much smaller diameter than the mouth of the collecting funnel. Thus, if the funnel exposes a surface of fifty square inches, and the measuring glass has a cross-section of one square inch, the fall of $\frac{1}{50}$ of an inch of rain on the funnel will give a quantity of water sufficient to fill the measuring glass to the depth of an inch. In this way the actual rainfall may be read to the thousandth part of an inch without trouble. The smallest diameter for a serviceable rain-gauge is five inches, and this size is well adapted for the traveller. A three-inch rain-gauge might be employed, but the results obtained with it are not so satisfactory. The rain-gauge should be placed in an open situation, so that it is not sheltered by any surrounding trees or buildings, and it ought to be firmly fixed by placing it between three wooden pegs driven securely into the ground. The mouth of the gauge should be level, and when the instrument is fixed, the rim of the funnel ought to be one foot above the ground. A spare measuring glass should be carried, but as there is always a considerable risk of breaking such fragile objects, it is well to carry also one or two small brass measures of the capacity of half an inch, two-tenths of an inch, and one-tenth of an inch of measured rainfall. In this way, although no satisfactory record could be kept of light rainfall, a very fair estimate may be made of any torrential showers, the half-inch measure being used first, and then the smaller measures, finally estimating by eye the fraction of the tenth of an inch that remains over. It must, however, be distinctly borne in mind that an estimate formed in this way is not an accurate measurement, and the fact of using the rough method must be stated in the note-book.

When snow falls along with rain, the melted snow is measured as equivalent to rainfall, and if the funnel of the rain-gauge should contain some unmelted snow at the time of observation, it should be warmed until the snow melts before a measurement is taken. When snow falls in a strong wind the drift that occurs makes it almost impossible to measure the amount accurately, but an effort should be made to estimate the average depth of the snow over a considerable area.

If the receiving bottle of the rain-gauge should be broken by frost or accident, any other bottle may be used, or in default of a bottle, the copper case itself will act as a receiver, although the risk of loss by evaporation, and by the wetting of a large surface in pouring out the water, is considerably increased.

At a fixed station the rain-gauge should be read every morning. The traveller who only exposes his rain-gauge during a halt should be careful to state the hours when it was exposed and when it was read.

Barometers.—The barometer is the most delicate, and at the same time the most important, instrument which a meteorologist has to employ. It requires particular care in transport, and must be very carefully mounted and read, while several accessory observations have to be made at each reading in order to ascertain the corrections required for the subsequent calculation of the results. The function of the barometer is to measure the pressure of the air at the time of observation, and this purpose may be carried out by the use of two different principles. The oldest and best method is to measure the height at which a column of heavy fluid is maintained in a tube entirely free from air. The weight of this column is equal to the weight of a column of the atmosphere of the same sectional area. Mercury being the densest fluid is the only one usually employed, because the column balancing a column of the atmosphere of equal sectional area is the shortest that can be obtained, and, consequently, a mercurial barometer is the most portable that can be constructed on this principle. The mercurial barometer has come to be recognised as the standard in all parts of the world.

The average height of the column of mercury in a barometer is about thirty inches, and, consequently, the whole instrument cannot well be made less than three feet long, so that when account is taken of the glass tube, and the amount of mercury it contains, it is long, fragile and heavy. To avoid the disadvantages inherent in such an instrument, the method of measuring the pressure of the air by the compression of a spring holding apart the sides of an air-free flexible metallic box was devised, and the aneroid barometer invented. The aneroid is graduated on the dial in "inches," *i.e.*, divisions each of which corresponds to a change of atmospheric pressure, equal to that measured by one inch of mercury in a standard barometer. Although a carefully constructed aneroid is a very useful instrument indeed, it is not to be trusted like a mercurial barometer kept in a proper place. But a good aneroid is likely to be much more serviceable to the ordinary traveller on the march than a standard mercurial barometer, every packing and unpacking of which exposes it to the risk of breakage, or to the equally fatal risk of air obtaining access to the vacuous space at the top of the tube.

We shall describe two types of mercurial barometer, one the Fortin, which is best adapted for use at a fixed station, the other devised by Prof. Collie and Capt. Deasy, which is portable enough for the use of travellers.

The Fortin Barometer.—The barometer must be kept in a room with as equable a temperature as possible; the instrument must be absolutely vertical—hence it should be hung freely and not touched while it is being read; it must be in a good light, and yet be sheltered from the direct rays of the sun. The measurement of the height of any mercurial barometer is that of the difference of level between the surface of the mercury in the tube and the surface of the mercury in the cistern. When the mercury rises in the tube it falls in the cistern, and *vice versâ*, although when the cistern is much wider than the tube the changes of level there are much less than those in the tube. In most barometers an arbitrary correction is made to allow for this change, the "inches" engraved on the scale not being true inches. In the Fortin barometer, however, the lower end of the measuring rod is brought in contact with the mercury in the cistern before every reading, and then the scale of inches engraved on the upper part of the measuring rod gives the true height of the column of mercury. In calculating the barometric pressure for purposes of comparison, five corrections have to be applied: (1) for temperature, which requires the temperature of the barometer at the time of reading to be observed, (2) for altitude, which necessitates knowing the elevation of the place of observation above sea-level, (3) for the force of gravity at sea-level, which requires the latitude to be known, (4) for the capillary attraction between the mercury and the glass tube, which is a constant for each barometer, (5) for the slight imperfection in engraving the scale (index error), which is also a constant for each instrument.

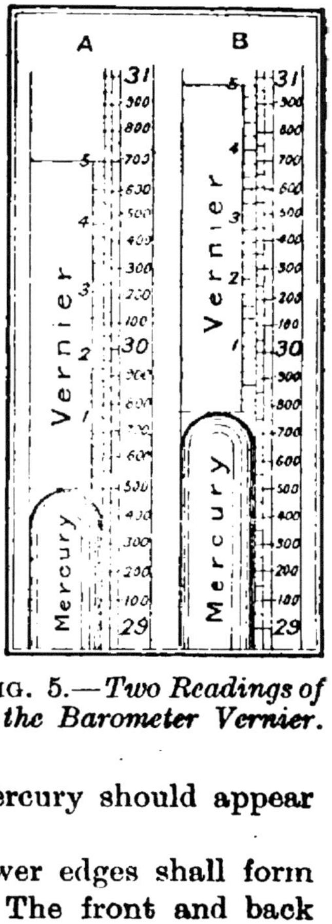

It is enough for the observer at a fixed station, and to such alone can the use of a Fortin barometer be recommended, to read the temperature on the thermometer attached to the barometer and to read the height of the mercury in the barometer tube. These two figures he is to enter in his note-book, and unless he is himself discussing the results, he should apply no correction whatever to them. The rules for observing, then, are :—

1. Read the attached thermometer and note the reading.

2. Bring the surface of the mercury in the cistern into contact with the ivory point which forms the extremity of the measuring rod by turning the screw at the bottom of the cistern. The ivory point and its reflected image in the mercury should appear just to touch each other and form a double cone.

Fig. 5.—*Two Readings of the Barometer Vernier.*

3. Adjust the vernier scale so that its two lower edges shall form a tangent to the *convex* surface of the mercury. The front and back edges of the vernier, the *top* of the mercury, and the eye of the observer are then in the same straight line.

4. Take the reading, and *enter the observation as read* without either correcting it to freezing point or reducing it to the sea-level.

The scale fixed to the barometer is divided into inches, tenths, and half-tenths, so that each division on this scale is equal to 0·050 inch.

The small movable scale or vernier attached to the instrument enables the observer to take more accurate readings; it is moved by a rack and pinion. Twenty-four spaces on the fixed scale correspond to twenty-five spaces on the vernier; hence each space on the fixed scale is larger than a space on the vernier by the twenty-fifth part of 0·050 inch, which is 0·002. Every long line on the vernier (marked 1, 2, 3, 4, and 5) thus corresponds to 0·010 inch. If the lower edge of the vernier coincides with a line on the fixed scale, and the upper edge with the twenty-fourth division of the latter higher up, the reading is at once supplied by the fixed scale as in A (Fig. 5), where it is 29·500 inches. If this coincidence does not take place, then read off the division on the fixed scale, above which the lower edge of the vernier stands. In B (Fig. 5) this is 29·750 inches. Next look along the vernier until one of its lines is found to coincide with a line on the fixed scale. In B this will be found to be the case with the second line above the figure "2." The reading of the barometer is therefore:—

On fixed scale	29·750
On vernier (12 × ·002)	·024
Correct reading	29·774

Should two lines on the vernier be in equally near agreement with two on the fixed scale, then the intermediate value should be adopted.

5. Lower the mercury in the cistern by turning the screw at the bottom until the surface is well below the ivory point; this is done to prevent the collection of impurities on the surface about the point.

The transport of barometers requires very great care in order to prevent the introduction of air into the tube or the fracture of the tube by the impact of the mercury against the top. To reduce the risk of these accidents, the barometer must be carried with the tube quite *full of mercury*, and in an inverted position, at least with the cistern end kept higher than the top of the tube. The flexible cistern of the Fortin type of barometer allows of it being screwed up tight so as to fill the tube and close the lower end of it. In case of breakage, the operation of fitting a new tube is not very difficult, but unless the tube has been carried out ready filled with mercury, this cannot well be attempted. In order to drive out the film of air adhering to the glass on the inside, it is necessary, after filling the tube, to raise its temperature to the boiling-point of mercury. No one should attempt either to fill or to change a barometer tube unless he has had practice in doing so under expert supervision beforehand.

The Collie Portable Mercurial Barometer.—This instrument is not likely to be broken in travelling. It is quickly set up, and from such tests as have been applied, it appears to give excellent results. The cistern and vacuum tube at the top are of equal diameter, and are connected by a flexible tube, and the difference in level of the mercury may be measured directly by means of a graduated rod, or as in Deasy's mounting by means of a vernier. There is no attached thermometer, but if the instrument be used in the open air, and is exposed for ten minutes or a quarter of an hour before using, it will be sufficient to note the temperature of the air in the usual way.

The upper end (Fig. 7) is about 2·5 inches long, and contains an air-trap, into which all the air that may accidentally enter the barometer, either by the tap leaking, through the rubber tubing, or through either of the joints, must find its way. The lower or reservoir end (Fig. 8) is about 4·5 inches long, and has an air-tight glass tap about an inch below the broad part. These ends are forced into the rubber tubing, and, as an additional precaution against leakage, copper wire is bound round the joints. The scale is cut on an aluminium bar, along which two carriages, to which the barometer is attached, move up and down, and they can be clamped to the bar at any place (Fig. 6). By means of the verniers attached to the carriages, which are divided to 0·002 of an inch, it is easy to estimate the height of the mercury to 0·001.

To use the barometer, the carriages are put on the scale bar; the lower one is clamped at the bottom of the bar, and the upper one some inches higher up; the barometer is attached to the carriages by clamps which fit over the joints; the rubber cap is removed from the reservoir end, the tap opened, the verniers put in the middle of their runs, and the upper carriage moved up the bar until there is a vacuum. By means of the screws on the right of the mercury, the verniers are moved up or down until the top of the mercury at each end is in line with the edges of the rings attached to the verniers, which fit round the glass ends. Both verniers are then read, and the difference gives the height of the barometer. The rubber cap on the reservoir end is merely to prevent the small quantity of mercury, which should be left above the tap when it

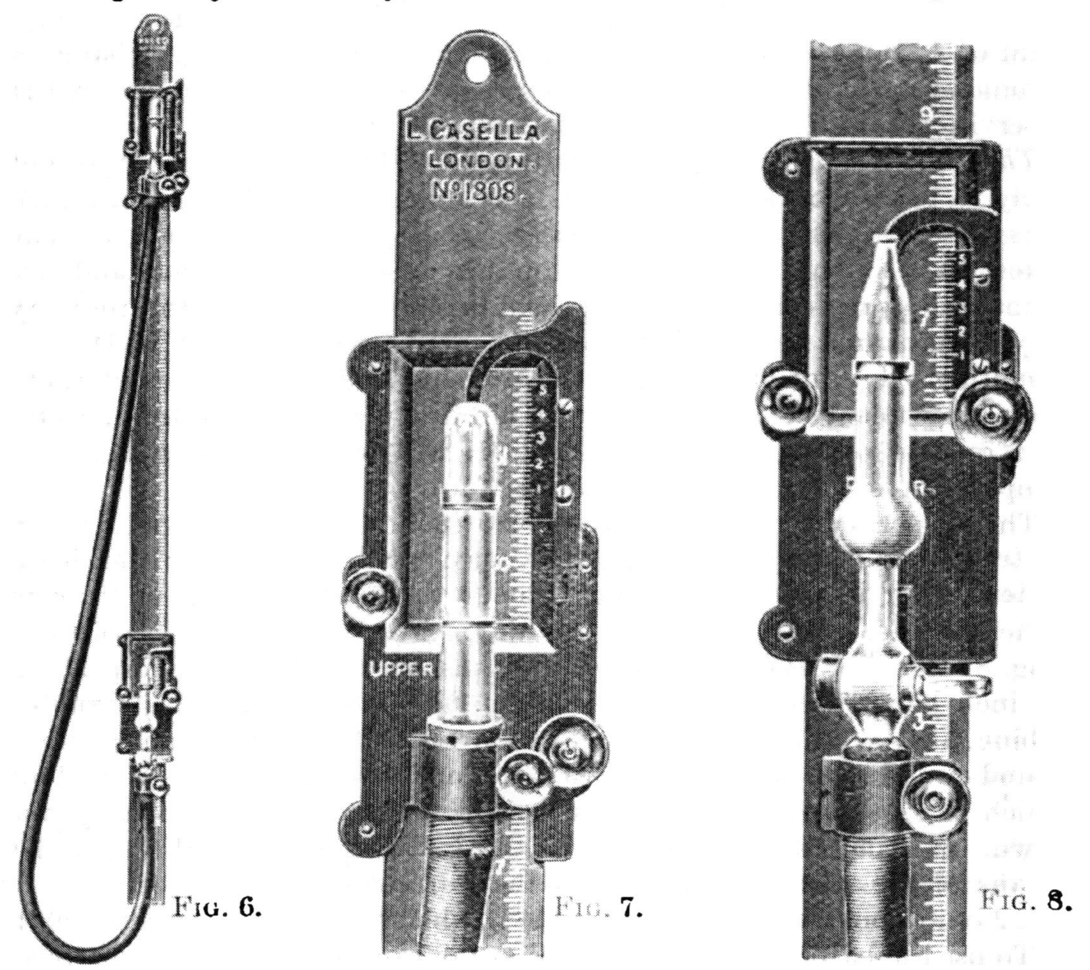

FIG. 6. THE COLLIE BAROMETER, WITH THE DEASY MOUNTING, IN ITS NORMAL WORKING POSITION.

FIG. 7. THE UPPER CARRIAGE AND VERNIER ON A LARGER SCALE, WITH BAROMETER ATTACHED.

FIG. 8. THE LOWER CARRIAGE AND VERNIER, WITH RESERVOIR END OF BAROMETER ATTACHED. (Same scale as Fig. 7.)

s closed, from being shaken out when travelling.

To pack up the barometer, lower the upper carriage *very slowly* until the mercury has touched the top of the glass; then detach the baromete from this carriage, and either let the upper end hang vertically below the reservoir, or detach the reservoir end from its carriage and raise it till the barometer hangs vertically. By this means the barometer is completely filled with mercury, and then the tap must be closed. The tube is then to be coiled away in its padded box. When too much air is found in the trap, it must be extracted by means of the air-pump.

The Aneroid Barometer.—The aneroid barometer is so convenient on account of its portability that, although much less trustworthy than a mercurial barometer, it is much more likely to be used by a traveller. Care should be taken in using it to see that the pointer has come to a position of equilibrium, and it should be tapped gently before reading. The eye must be brought directly over the end of the pointer, and the reading made to one-hundredth of an inch, the barometer being held in a horizontal position. Every opportunity of comparing the aneroid with a standard mercurial barometer should be taken, and a note made of the readings of both. The mercurial barometer will require to be corrected for temperature before its indications can be used for correcting the aneroid, as all good aneroids are compensated for changes of temperature. The readings of an aneroid give a very fair idea of the changes of atmospheric pressure, and are very much better than none at all, although they cannot in any case be accepted as of the highest order of accuracy.

The new Watkin mountain aneroid, which is so constructed as to be thrown into gear at the moment when it is read, appears to be free from the worst errors of the ordinary aneroid.

For climatological purposes, it is impossible to make barometric observations of value while travelling unless the altitude of each camping-place is accurately known. This is practically never the case except when travelling along the sea-shore or the margin of a great lake the elevation of which has been determined. But, meteorology apart, barometric readings in any little known country are of value, because by comparing them with simultaneous readings taken at a neighbouring fixed station, new data as to the altitude of the country may be obtained. While in camp, it would be an extremely useful thing to make barometer readings, even with an aneroid, every two hours, in order to get some information as to the normal daily range of atmospheric pressure.

The Boiling-point Thermometer.—The temperature at which water boils depends on the pressure of the atmosphere, so that an accurate observation of the boiling-point of water enables the pressure of the atmosphere at the moment of observation to be determined with the utmost accuracy. This method of determining atmospheric pressure having been used hitherto almost solely for the purpose of measuring altitudes, the boiling-point thermometer is usually known as the Hypsometer, but its records are quite as valuable for use at fixed stations as in mountain climbing. Mr. J. Y. Buchanan recommends the use of a boiling-point thermometer with a very open scale graduated to fiftieths of a degree Centigrade and entirely enclosed in a wide glass tube through which steam from water boiling in a copper vessel is passing. On a thermometer of this kind change of pressure can be measured by the change of boiling-point more accurately than with the aid of a mercurial barometer. See Table, Vol. I., p. 293.

2. Observations for Forecasting the Weather.—The familiar name of "weather-glass" is appropriately applied to the barometer, for in most parts of the world it is the surest indicator of any approaching storm.

The scientific prediction of the weather by means of the barometer involves the comparison of the simultaneous readings of barometers over as wide an area as possible, and can only be carried out where there is a complete telegraph system and a public department charged with the work. The storms of wind and rain which break the more usual steady weather are always associated with the formation of centres of low atmospheric pressure towards which wind blows in from every side. These atmospheric depressions move, as a rule, in fairly regular tracts, the rate of movement of the centre of the depression having no relation to the rate at which the wind blows or to the direction of the wind. The term *cyclone* is usually applied to such a moving depression, because of the rotating winds round the centre; but the size of a cyclone may vary from a vast atmospheric eddy extending across the whole breadth of the Atlantic to one only a few miles in diameter. The strength of the wind in a cyclone depends on the barometric gradient; in other words, the greater the difference in atmospheric pressure between two neighbouring points the stronger is the wind that blows between them. Or, when a cyclone is passing over an observer, the more rapidly the barometer falls *or rises* the stronger may the wind be expected to blow.

In direct contrast to the cyclone or depression is the system of high pressure rising to a centre from which the wind blows out on every side.

This is called an anticyclone, and is a condition which, once established, may last for many days, or even weeks, without change. It is the typical condition for dry calm weather in all parts of the world.

The direction of movement of the centres of cyclones in the northern hemisphere is usually westward and northward near the equator, the path of the centre bending to the right as it proceeds, and becoming ultimately eastward and southward. In the southern hemisphere the direction of the centre near the equator is westward and southward, turning towards the left as it proceeds. The rotation of the wind about the centre of a cyclone in the northern hemisphere is inwards towards the centre in the direction opposite to the hands of a watch, and in the southern hemisphere it is in the direction in which the hands of a watch move (Fig. 9). In the centre of a cyclone there is a calm, a well-known danger to sailing ships caught in such a storm at sea, because there is no wind to move the vessel, but a tremendous sea driven in from the gale which rages all round from every point of the compass. The law of storms has been very fully studied, and rules have been drawn up to enable sailors to ascertain the direction in which the centre of an approaching cyclone lies and the direction in which it is moving. In a work intended mainly for travellers on land it is not necessary to give these rules; all that is required is to tell how the approach of dangerous storms may be ascertained some time in advance. The fact that the barometer is high or low is in itself of no value for prediction. The important thing to know is the distribution of atmospheric pressure at a given moment over a considerable area.

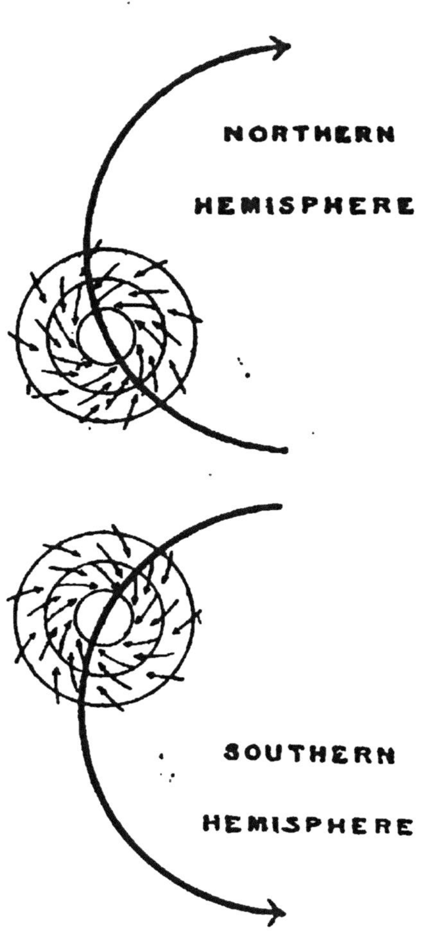

FIG. 9.—CYCLONE PATHS AND CIRCULATION OF WINDS IN CYCLONES IN THE NORTHERN AND SOUTHERN HEMISPHERES.

To the isolated observer this is impossible, and he can only judge of the state of the atmosphere by observing the *rate* at which the barometer is falling or rising. Thus, if for several days the barometer has been steadily and slowly rising, he will probably be right in believing that an anticyclonic condition is establishing itself, and that the weather may be expected to continue fine for many days to come, even after a gradual fall of the barometer begins. A sudden fall of the barometer, on the other hand, is always a sign of wind, and usually of wet weather as well. This is a particularly valuable sign of approaching storm in those parts of the world where, as in the tropics, the normal weather is very uniform and steady. In such places a very sudden fall, say one-tenth of an inch in an hour, is a sure precursor of a violent storm. As the barometer continues to fall, the wind will probably continue to increase in force, and when the barometer reaches its lowest point it will either fall calm (if the centre of the cyclone is passing over the observer) or suddenly change in direction. The rapid rise of the barometer after a great depression is also always accompanied by strong wind, though not so frequently by rain.

It must be clearly understood that these remarks refer only to observations at a fixed station. If a fall of barometer is observed in travelling, it may be due either to a change in the state of the atmosphere or to a change in the traveller's height above sea-level. This is the reason why it is absolutely essential, in making barometric estimates of height (or boiling-point determinations), to have simultaneous observations going on at a base-station, or preferably at a series of intermediate stations.

The ordinary prognostics of the approach of rain or bad weather differ in different localities, and require a considerable amount of local knowledge before they can be utilised. The peculiar absorption band in the solar spectrum due to the water vapour of the Earth's atmosphere, and called the rain-band, is a valuable guide to an experienced observer with a spectroscope in predicting rain. The only instrument, however, likely to be useful to the ordinary traveller is the wet and dry bulb thermometer. When the two thermometers have the same reading, indicating saturation of water vapour, or when they approach at temperatures above 60° F. within two degrees or so, rain may be expected, or possibly mist. The appearance of low clouds clinging to the hillsides is an indication that the temperature at the place where they are is below the dew-point. The appearance of the upper clouds, taken in conjunction with the readings of the barometer, is a valuable indication of forthcoming weather changes. The increase of cirrus clouds in a clear sky with a falling barometer, or the appearance of a solar or lunar halo, may be taken as a sure sign of an approaching cyclone, the intensity of which may be foreseen by the *rate* at which the barometer is falling.

While the weather of places on the west coasts of temperate continents exposed to the prevailing sea-wind is usually made up of a succession of cyclones of different degrees of intensity, and of the anticyclonic intervals between them, over the greater part of the Earth's surface the climate is much more uniform, and the seasonal changes are the principal cause of changes of weather. To understand these general conditions it is necessary to consider the elements of climatology.

8. Outlines of Climatology.—The air is in constant movement on account of the unequal way in which the heat of the sun falls on different parts of the Earth's surface, and at different seasons of the year. All the conditions of the atmosphere show a certain diurnal periodicity which is most marked in the regions of steady climate between and near the tropics. Thus, as a rule, the minimum temperature of the air occurs just before sunrise, the maximum temperature from two to three hours after noon. The amount of difference between the maximum and minimum temperature of the day (daily range) is least near the sea or in wet regions (a maritime climate) and greatest in the interior of the continents, especially where the rainfall is slight (a continental climate). Over the sea itself the daily range of air temperature averages only 3 Fahrenheit degrees; but in the heart of a continent, especially in a desert, it may exceed 60 Fahrenheit degrees.

Diurnal changes of pressure are proportionally much smaller in amount than changes of temperature, and are to be observed as a regular phenomenon only in the tropics, or elsewhere during very settled weather. There are usually two maxima daily, abont 10 A.M. and 10 P.M., and two minima occurring about 4 A.M. and 4 P.M. It is only in rare cases that the total barometric range exceeds 0·10 inch, very frequently it is not greater than 0·04 inch. Still it is convenient to remember in the tropics that a fall of the barometer not greater than 0·10 inch between 10 A.M. and 4 P.M. is to be expected, and does not indicate either the approach of a storm (if the observer is at rest) or the ascent of 100 feet (if he is on the march).

Associated with the diurnal changes of temperature in settled weather are changes of wind due to local configuration of the ground. The wind, for example, usually blows up a mountain side, or up a steep valley, during the day, and down a mountain, or down a steep valley, during the night. So, too, the regular land and sea breezes found on the borders of the sea or of great lakes blow from water to land in the day time and from land to water at night. Here the determining cause is the fact that land is warmed and cooled by radiation, and in turn heats or chills the air much more than water does. In the settled climates of high tropical plateaus a regular diurnal change of wind direction has been observed, the wind blowing successively from all points of the compass.

Similar diurnal periodicities occur in the amount of cloud, in the moisture of the air, the fall of rain, the occurrence of thunderstorms, etc. It is also to be noticed in the flow of rivers in mountainous regions where the streams take their rise from glaciers or snow, the rapid melting of which by the heat of the sun causes the volume of water to increase greatly in the afternoon, while the cessation or reduction of the rate of melting at night diminishes the volume of the river in the morning and forenoon.

Periodic changes of greater amount but similar in kind are produced by the alternation of the seasons, the difference between the mean values of the months in which the phenomena are at a maximum and minimum respectively being termed the annual range. With regard to temperature, very moderate changes occur in the tropical zones where the altitude of the noon-day sun is always great, and the length of day and night varies little with the season (for the most part less than 5 Fahrenheit degrees); but in the temperate and frigid zones there are strongly marked annual changes. As in the case of daily range, proximity to the sea is a controlling factor in the annual range of temperature. To take a very characteristic instance, the annual range between the mean temperature of July and January is about 23 Fahrenheit degrees in the Lofoten islands on the margin of the Atlantic, while it is 120 Fahrenheit degrees at Verkhoyansk in the same latitude, but in the centre of the Asiatic continent.

The extreme months for air temperature are January and July in almost every part of the world, the maximum occurring north of the equatorial belt in July and south of it in January.

The annual changes in barometric pressure and wind are equally marked. The belt of low pressure which lies nearly under the vertical sun moves northward over the surface of the globe in the northern summer, coming to its most northerly position in July: returning southward after the sun, it reaches its most southerly position in January. This belt of low pressure is also a belt of calms, known by sailors as the Doldrums, and it is a belt of frequent rains, so that as it approaches and passes over a place there is a rainy season, followed by a dry season when it retires. Near the mean position of the belt of low pressure, where it passes over a place twice in the year, there are two rainy seasons. The low pressure belt is bordered to north and south by belts of high atmospheric pressure, from which the trade-winds blow towards the equator, and the westerly anti-trades blow towards the poles. These are also subject to the annual change; but the different action of land and sea on the distribution of pressure exercises a greater influence than does the difference of latitude. As the greater heating and cooling of the land each day causes the phenomena of daily land and sea breezes, so the greater heating and cooling of the land between summer and winter causes seasonal land and sea winds, blowing from land to sea in winter, from sea to land in summer. Generally speaking, the pressure is greater—in the same latitude—where the air is cooler, so that outside the frigid zones cold areas are usually areas of high pressure, from which wind blows out in every direction, while warm areas are areas of low pressure towards which wind blows in on every side.

The distribution of rainfall on the land is dependent on the direction of the rain-bringing wind and the configuration of the surface. Thus when the rain-bringing wind meets a mountain range, it deposits a great rainfall on the exposed slopes, but passes over as a dry wind which yields little rain to the region beyond. In places where the wind changes with the season, as in southern Asia, the distribution of rainfall is entirely different during the continuance of the different monsoons.

All these questions of normal climate can be more easily illustrated on maps than explained by words. But the reader must be cautioned against taking the condensed and generalised representations of small-scale maps as showing all that is known on the subject. Even the magnificent plates in the 'Atlas of Meteorology,' which forms part of Bartholomew's Physical Atlas, cannot show everything that is known; and in many parts of the world so little has yet been ascertained as to the climatic conditions that generations of observers will be required to make it possible for meteorologists to draw a uniform trustworthy map of the whole world showing the distribution of any one element of climate.

Isothermal maps.—The principle of an isothermal map is that of representing the distribution of temperature by drawing lines through all the places where the temperature is the same at a given time. It is usual to take this time as an average month in an average year. Thus in a map of isotherms for January (see p. 50), what is shown is not the temperature of any particular day in any particular January, but that of an average day in a long series of Januaries. Hence it is not likely that the *exact* distribution of temperature shown in the map will ever be found on any January day; but it is to be expected that most days in every January will have a distribution of temperature which is very similar to that shown. The same is of course true of maps showing pressure, or rainfall, or any other average condition.

Again, the isotherm is necessarily constructed from average temperatures which have been corrected so as to be applicable to the same level. On the equator, for instance, the summit of a lofty mountain is seen by the snow on it to have a temperature not exceeding 32° F., while at sea-level the temperature may be 90°. But observations have been made showing the rate at which the temperature of the air diminishes as the height increases, and although the rate varies in different places and at different seasons, it may be taken as about one Fahrenheit degree in 300 feet. Now if the mountain top with a temperature of say 30° F. is known to be 18,000 feet above the sea, the addition of 1° for every 300 feet, or 60° altogether, would give the temperature of 90° as that corresponding to sea-level. By applying such corrections, the isothermal maps have been constructed to show the distribution of temperature at the level of the sea. In order to compare the temperature he has observed with that on the map the observer must calculate the average of his daily observations for the month in question, and then make the correction for the altitude of his station.

Similarly, in ascertaining from an isothermal map the mean temperature of a particular place, care must be taken to subtract from the number of degrees of the isotherm passing through the place one degree for every 300 feet of elevation. Of course it will usually happen that no isotherm as shown on the map runs through the point the mean temperature of which it is desired to obtain. In that case the temperature at the point will be found by considering its relative position between the two nearest isotherms. Thus, if it lie half-way between the lines of 60° and 70°—measured perpendicularly to the isotherms—the temperature of 65° may be assumed; if it lies one-tenth of the distance from 60° and nine-tenths from 70°, it is safe to assume 61°; if three-tenths from 70° and seven-tenths from 60°, then assume 67°. If the point lie in a loop of a single isotherm, *e.g.*, Cape St. Roque, the eastern point of South America in the map for January, lying within the 80° isotherm, one can only guess that the temperature is above 80° and it may be assumed to be below 85°. The method of representation is unsatisfactory in such a case.

These facts being borne in mind, the study of isotherm maps will be found to give an excellent general idea of the distribution of climate at sea-level, and if the contour lines of 600 and 6000 feet are traced on the maps the areas within which corrections of over −2° and −20° have to be applied to the isothermal values to get the temperature at the place will be easily recognised.

Isobaric Maps.—Isobars are drawn from the data of the height of the barometer corrected to sea-level values and to the temperature of 32° F., exactly in the same way as isotherms are drawn from the data of thermometer readings or contour lines from data of elevation measurements. The practical value of the study of isobars is very great, because of the importance of assuming a probable value of sea-level pressure in reducing the barometric or boiling-point thermometer readings for determining elevation, and also because of the intimate relation between the form and proximity of isobars and the direction and force of the winds.

Barometric gradient is measured by the difference between the isobars per unit of length. For instance, gradient is frequently expressed in the number of hundredths of an inch difference between barometers fifteen nautical miles apart. The greater the gradient of pressure is, the more closely together must the isobars be drawn in order to represent it. For example, in the isobaric map for January (p. 50) a very steep gradient is shown on the east coast of Asia, north of Japan, and a remarkably gentle gradient in the interior of Asia from the Black Sea eastward. The steeper the gradient the stronger is the wind.

The arrows in the isobaric maps (which are represented flying with the wind) show the average directions of the wind over the world for the months in question. The relation they bear to the isobars becomes clear on inspection, although, on account of the greater number of observations available for some parts of the world than for others, all the arrows are not drawn with the same amount of certainty, and the direction of a few contradicts that of most. As a general rule, the following facts may be taken as absolutely established: (1) Wherever there is a region of high pressure the wind blows out from it in all directions. (2) Wherever there is a region of low pressure the wind blows in towards it from every side. (3) The wind never blows perpendicularly to the isobars or directly from higher to lower pressure, but always in a curved or spiral path inclined to the isobars. (4) In the northern hemisphere the wind blows out from a high pressure area in the same direction as the hands of a watch move, but in the southern hemisphere in the opposite direction. Also in the northern hemisphere the wind blows into a low-pressure area in the direction opposite to that of the hands of a watch and in the southern hemisphere in the same direction as the hands of a watch move. (5) Recognising that the wind blows nearly parallel to the direction of the isobars, the following statement (known as Buys Ballot's Law) expresses its direction both for high-pressure and for low-pressure areas: If you stand with the lower pressure on your left hand, and the higher pressure on your right hand, in the northern hemisphere the wind will be blowing on your back, but in the southern hemisphere in your face.

Rainfall Maps.—Rainfall is represented on maps by lines of equal precipitation termed Isohyets. These represent actual figures without reduction for elevation or other local conditions, and a rainfall map can consequently be studied as a direct record of observed facts. The map (p. 50) of mean annual rainfall brings out clearly the equatorial zone of heavy rains crossing the Amazon valley, the Congo valley, the south-eastern peninsulas of Asia and the Malay archipelago. North and south of this belt are the nearly rainless regions of the tropical deserts, extended northward and southward over the continents, and merging nearer the poles into the fairly-watered temperate zones. The rainfall maps for separate months show the intimate relation between rainfall and the direction of the wind taken in conjunction with the configuration of the land. Even on the coast, when the prevailing wind is off shore, there may be scarcely any rain, as on the west coast of tropical South America. In the very heart of a continent the rainfall may be very heavy where the sea-wind blows across a great plain before striking the mountains, as is illustrated by the eastern slope of the Andes. Rainfall is, however, one of the most inconstant elements of meteorology, and the actual rainfall of any year may differ very widely from the average. The practical value of exact statistics of rainfall is, however, greater than that of any other climatological condition; for the water supply and the fertility of the land depend in every case on the rain that falls either locally or on the heights of the water-sheds.

In order to pursue the subject further the chapters on the atmosphere in the writer's 'Realm of Nature' (London, Murray; New York, Scribner) or the little treatise on 'Meteorology' by Mr. H. N. Dickson (London, Methuen) will be found useful and compact. Dr. R. H. Scott's 'Meteorology' and Mr. R. Abercromby's 'Weather' in the International Science Series (London, Kegan Paul), and in a special degree Prof. W. M. Davis' 'Elementary Meteorology' (Boston), will be found of great value. The most systematic treatment of climatology will be found in Hann's 'Handbuch der Klimatologie,' 3 vols. (Stuttgart), which contains numerous references to special works; the essential part of this treatise is translated by Professor R. de C. Ward, under the title of 'Handbook of Climatology,' Part I. (Macmillan). The most important work of all is the great 'Atlas of Meteorology' by Dr. A. Buchan and Dr. A. J. Herbertson, forming Vol. III. of Bartholomew's 'Atlas of Physical Geography' (London, Constable), which gives an unrivalled series of climate and weather-maps with explanatory letterpress and appendices containing complete bibliographies and lists of Weather Services in all parts of the world.

Daily synoptical weather-maps are published by the Weather Service of almost every civilised country. That for the United Kingdom may be obtained for a subscription of £1 per annum from the Secretary of the Meteorological Office, 63, Victoria Street, London, S.W. The only weather-maps of large areas produced regularly are the Pilot Charts of the North Atlantic and the North Pacific, published monthly by the Hydrographic Office at Washington, and those of the North Atlantic and Mediterranean published monthly by the Meteorological Office in London. These show the tracks of cyclones, and give a great deal of information as to the meteorology and currents of the oceans. They are intended primarily for the use of sailors.

The following list gives the name of the official weather service of all countries outside Europe and the town in which the head office is situated. Application might be made to any of these offices for information as to the stations where standard instruments are established in the country in question. The list is taken from the fuller list of stations published in the 'Atlas of Meteorology.'

EXTRA-EUROPEAN WEATHER SERVICES.

Country.	Name.	Place.
America :—		
Canada	Meteorological Service of the Dominion of Canada	Toronto.
United States and West Indies ..	Weather Bureau	Washington.
Mexico	Observatorio meteorologico magnetico central	Mexico.
Salvador	Observatorio Astronomico y Meteorologico	San Salvador.
Guatemala	Instituto fisico-geografico	San José.
Cuba	Real Colegio de Belen de la compania de Jesus	Havana.
Brazil	Central Meteorological Department of the Navy	Rio de Janeiro.
Uruguay	Observatorio meteorologico del colegio pio de Villa Colon	Montevideo.
Argentina	Oficina meteorologica Argentina ..	Cordoba.
Chile	Oficina meteorologica	Santiago.
Peru	Harvard College Observatory	Arequipa.
Africa :—		
Algeria	Service météorologique	Algiers.
British Central Africa	.. ..	Zomba.
Cape Colony ..	Meteorological Commission	Cape Town.
Madagascar ..	Observatoire	Tananarive.
Mauritius	Royal Alfred Observatory	Pamplemousses.
Natal	Observatory	Durban.
Transvaal	Meteorological Observatory	Johannesburg.
Orange River Colony	Gray's College	Bloemfontein.
Asia :—		
India	Meteorological Department	Calcutta.
Ceylon	Surveyor-General's Office	Colombo.
Hongkong	Observatory	Hongkong.
China	Imperial Maritime Customs	Peking.
,, ...	Observatoire de Zi-ka-wei	Shanghai.
Philippine Islands	Weather Bureau	Manila.
Dutch East Indies	Magnetic and Meteorological Observatory	Batavia.

EXTRA-EUROPEAN WEATHER SERVICES—*continued.*

Country.	Name.	Place.
Australasia and Pacific:—		
New South Wales	Meteorological Service	Sydney.
Queensland	Meteorological Service	Brisbane.
Victoria	Meteorological Service	Melbourne.
South Australia ..	Meteorological Service	Adelaide.
Western Australia	Meteorological Service	Perth.
Tasmania	Government Meteorologist	Hobart.
New Zealand ..	Meteorological Office	Wellington.
Hawaii	Weather Bureau	Honolulu.

In addition to the above, which are regular government services specially organised for publishing and utilising the data from numerous observing stations, there are many isolated stations in all parts of the world. There are no colonies which do not possess some meteorological stations, and at many mission stations meteorological observations are made. It would always be well for a traveller to try to ascertain where in the vicinity of his route meteorological stations have been established and over what period of time their records extend.

In the Table, which is taken from Marriott's "Hints to Meteorological Observers," is given the relative humidity for every 2° of temperature from 20° to 80°, and for every two-tenths of a degree of difference between the dry and wet-bulb readings from 0°·2 to 18°·0.

To use the Table: Look in the column on the left or right for the nearest degree to the dry-bulb reading; then carry the eye horizontally along until the column is reached corresponding to the difference between the readings of the dry and wet-bulb thermometers, when the relative humidity will be found. Intermediate readings can be interpolated in the usual way.

Example: Dry-bulb 58°·5, wet-bulb 52°·7, the difference is 5°·8. Having found 58° in the column on the left or right, run the eye along this line until the column under 5°·3 is reached, when the relative humidity will be found, viz., 67.

TABLE OF

Dry Bulb Reading.	DIFFERENCE BETWEEN THE READINGS OF													
	° 0·2	° 0·4	° 0·6	° 0·8	° 1·0	° 1·2	° 1·4	° 1·6	° 1·8	° 2·0	° 2·2	° 2·4	° 2·6	° 2·8
°														
20	93	86	80	74	68	63	59	55	51	47	44	41	37	34
22	94	88	82	76	70	65	60	56	53	50	47	44	41	38
24	94	88	83	78	73	69	65	61	57	53	50	47	44	42
26	95	90	85	80	76	72	68	64	61	57	54	51	48	45
28	96	92	88	84	80	76	72	69	66	63	60	57	54	52
30	96	93	90	87	84	81	78	75	72	70	67	64	62	60
32	97	94	91	89	87	84	82	80	78	76	73	71	69	67
34	97	95	93	91	89	87	85	83	81	80	78	76	74	72
36	98	96	94	93	91	89	87	86	84	82	80	79	77	76
38	98	96	94	93	91	89	87	86	85	83	81	80	78	77
40	98	97	95	94	92	90	88	87	86	84	82	81	79	78
42	98	97	95	94	92	90	88	87	86	84	83	81	80	78
44	98	97	95	94	92	90	88	87	86	84	83	82	81	79
46	99	97	95	94	92	91	89	88	87	85	84	82	81	79
48	99	97	95	94	92	91	89	88	87	85	84	82	81	80
50	99	97	96	94	93	92	90	89	88	86	85	83	82	80
52	99	97	96	94	93	92	90	89	88	86	85	83	82	81
54	99	97	96	94	93	92	90	89	88	86	85	83	82	81
56	99	97	96	94	93	92	90	89	88	87	86	84	83	82
58	99	97	96	94	93	92	90	89	88	87	86	85	84	83
60	99	97	96	94	93	92	91	90	89	88	86	85	84	83
62	99	98	96	95	94	93	91	90	89	88	86	85	84	83
64	99	98	96	95	94	93	91	90	89	88	86	85	84	83
66	99	98	96	95	94	93	91	90	89	88	87	86	85	84
68	99	98	96	95	94	93	92	91	90	89	87	86	85	84
70	99	98	96	95	94	93	92	91	90	89	87	86	85	84
72	99	98	96	95	94	93	92	91	90	89	87	86	85	84
74	99	98	96	95	94	93	92	91	90	89	88	87	86	85
76	99	98	97	96	95	93	92	91	90	89	88	87	86	85
78	99	98	97	96	95	94	93	92	91	90	89	88	87	86
80	99	98	97	96	95	94	93	92	91	90	89	88	87	85

RELATIVE HUMIDITY.

THE DRY AND WET BULB THERMOMETERS.															Dry Bulb Reading.
° 3·0	° 3·2	° 3·4	° 3·6	° 3·8	° 4·0	° 4·2	° 4·4	° 4·6	° 4·8	° 5·0	° 5·2	° 5·4	° 5·6	° 5·8	°
32	29	27	25	23	21	19	18	17	16	15	14	13	12	11	20
36	33	31	29	27	25	23	21	19	18	17	15	14	13	12	22
39	36	34	32	30	29	27	25	23	21	19	18	17	16	15	24
43	41	39	37	35	33	31	29	27	26	25	23	21	20	19	26
50	48	46	44	42	40	38	36	34	33	32	30	29	27	26	28
58	55	53	51	49	47	45	44	42	40	39	37	36	35	33	30
65	63	61	60	58	[illegible]	54	53	51	50	48	47	45	44	42	32
71	69	67	65	63	62	60	59	57	56	55	53	52	51	50	34
74	72	71	69	68	66	64	63	61	60	59	58	57	55	54	36
76	74	73	71	70	69	67	66	64	63	62	61	60	58	57	38
76	75	74	72	71	70	68	67	65	64	63	62	61	60	59	40
77	76	75	73	72	70	69	68	66	65	64	63	62	61	60	42
78	77	76	74	73	72	71	70	68	67	65	64	63	62	61	44
78	77	76	74	73	72	71	70	68	67	66	65	64	63	62	46
79	78	77	75	74	73	72	71	69	68	67	66	65	64	63	48
79	78	77	75	74	73	72	71	70	69	68	66	65	64	63	50
80	79	78	7[illegible]	75	74	73	72	71	70	69	67	66	65	64	52
80	79	78	77	76	75	73	72	71	70	69	68	67	[illegible]	65	54
81	80	79	77	7[illegible]	75	74	73	72	71	70	69	68	67	66	56
82	81	80	78	77	76	75	74	73	72	71	70	69	68	67	58
82	81	80	78	77	76	75	74	73	72	71	70	69	68	67	60
82	81	80	79	78	77	76	75	74	73	72	71	70	69	68	62
82	81	80	79	78	77	76	75	74	73	72	71	70	69	68	64
83	82	81	80	79	78	77	76	75	74	73	72	71	70	69	66
83	82	81	80	79	78	77	76	75	74	73	72	71	70	69	68
83	82	81	80	79	78	77	76	75	74	74	73	72	71	70	70
83	82	81	80	79	79	78	77	76	75	74	73	72	71	70	72
84	83	82	81	80	79	78	77	76	75	74	73	72	71	70	74
84	83	82	81	80	80	79	78	77	76	75	74	73	72	71	76
85	84	83	82	81	80	79	78	77	76	75	74	73	72	71	78
85	84	83	82	81	80	79	78	77	76	76	75	74	73	72	80

TABLE OF RELATIVE

Dry Bulb Reading.	DIFFERENCE BETWEEN THE READINGS OF														
°	6·0°	6·2°	6·4°	6·6°	6·8°	7·0°	7·2°	7·4°	7·6°	7·8°	8·0°	8·2°	8·4°	8·6°	8·8°
20	10	9	8	7	6	6	6	5	5	5	5	..	..	..	..
22	11	10	9	9	8	8	7	7	6	6	5	5	5	5	4
24	14	13	12	11	10	10	9	9	8	8	7	7	6	6	6
26	18	17	16	15	14	14	13	12	11	10	10	9	9	8	8
28	25	24	23	21	20	19	18	17	16	15	15	14	14	13	13
30	32	31	30	29	28	28	27	26	25	24	23	22	21	20	19
32	41	39	38	37	36	35	34	33	32	31	30	29	29	28	27
34	49	47	46	45	44	43	41	40	39	38	37	36	36	35	34
36	53	52	51	49	48	47	46	45	44	43	42	41	40	39	39
38	56	55	54	52	51	50	49	48	47	46	45	44	43	42	42
40	58	57	56	54	53	52	51	50	49	48	47	46	45	44	44
42	59	58	57	56	55	54	53	52	51	50	49	48	47	46	45
44	60	59	58	57	56	55	54	53	52	51	50	49	48	47	46
46	61	60	59	58	57	56	55	54	53	52	51	50	49	48	47
48	62	61	60	59	58	57	56	55	54	53	52	51	50	49	48
50	62	61	60	59	58	58	57	56	55	54	53	52	51	50	49
52	63	62	61	60	59	59	58	57	56	55	54	53	52	51	50
54	64	63	62	61	60	59	58	57	56	55	55	54	53	52	51
56	65	64	63	62	61	60	59	58	57	56	56	55	54	53	52
58	66	65	64	63	62	61	60	59	58	57	57	56	55	54	53
60	66	65	64	63	62	62	61	60	59	59	58	57	56	55	54
62	67	66	65	64	63	63	62	61	60	59	58	57	56	56	55
64	68	67	66	65	64	63	62	61	60	59	59	58	57	57	56
66	68	67	66	65	64	64	63	62	61	60	60	59	58	58	57
68	69	68	67	66	65	65	64	63	62	61	60	59	58	58	57
70	69	68	67	66	65	65	64	63	62	61	61	60	59	59	58
72	70	69	68	67	66	65	64	63	62	61	61	60	60	59	58
74	70	69	68	67	66	66	65	64	63	62	62	61	61	60	59
76	71	70	69	68	67	66	65	65	64	63	63	62	61	61	60
78	71	70	69	68	67	67	66	65	64	63	63	62	61	61	60
80	72	71	70	69	68	67	67	66	65	64	64	63	62	62	61

HUMIDITY.—*Continued.*

THE DRY AND WET BULB THERMOMETERS.															Dry Bulb Reading.
° 9·0	° 9·2	° 9·4	° 9·6	° 9·8	° 10·0	° 10·2	° 10·4	° 10·6	° 10·8	° 11·0	° 11·2	° 11·4	° 11·6	° 11·8	°
..	..	..	..	..	..	..	..	..	..	..	..	..	..	..	20
4	..	..	..	..	..	..	..	..	..	..	..	..	..	..	22
5	5	5	5	5	4	..	..	..	..	..	..	..	..	..	24
7	7	6	6	6	5	5	5	5	5	4	..	..	..	..	26
12	12	11	10	10	9	9	8	8	7	7	7	6	6	6	28
18	17	16	15	15	14	14	13	13	12	12	12	11	11	10	30
27	26	25	24	23	23	22	21	20	20	19	19	18	17	17	32
33	33	32	31	30	30	29	28	27	27	26	26	25	24	24	34
38	37	36	35	34	34	33	32	31	31	30	30	29	28	28	36
41	40	39	38	37	36	35	34	33	33	32	32	31	30	30	38
43	42	41	40	39	38	37	36	35	35	34	34	33	32	32	40
44	43	42	41	40	40	39	38	37	36	36	35	34	33	32	42
46	45	44	43	42	41	41	40	39	39	38	37	36	35	34	44
47	46	45	44	43	43	42	41	40	39	39	38	37	36	35	46
48	47	46	45	44	44	43	42	41	40	40	39	38	37	36	48
49	48	47	46	45	45	44	43	42	41	41	40	40	39	39	50
50	49	48	47	46	46	45	44	44	43	43	42	41	40	40	52
51	50	49	48	47	47	46	45	45	44	44	43	42	41	41	54
52	51	50	49	48	48	47	46	46	45	45	44	43	42	42	56
53	52	51	50	49	49	48	47	47	46	46	45	45	44	44	58
54	53	52	51	50	50	49	48	47	46	46	45	45	44	44	60
55	54	53	52	51	51	50	49	48	47	47	46	46	45	45	62
55	55	54	53	52	52	51	50	49	48	48	47	47	46	46	64
56	55	54	53	52	52	51	51	50	49	49	48	48	47	47	66
56	56	55	54	53	53	52	52	51	50	50	49	48	47	47	68
57	57	56	55	54	54	53	52	51	50	50	49	49	48	48	70
58	57	56	55	54	54	53	53	52	51	51	50	50	49	49	72
59	58	57	56	55	55	54	54	53	52	52	51	50	49	49	74
59	59	58	57	56	56	55	54	53	52	52	51	51	50	50	76
60	59	58	57	56	56	55	55	54	53	53	52	52	51	51	78
60	60	59	58	57	57	56	55	54	53	53	52	52	51	51	80

TABLE OF RELATIVE

Dry Bulb Reading.	Difference between the Readings of														
°	12·0°	12·2°	12·4°	12·6°	12·8°	13·0°	13·2°	13·4°	13·6°	13·8°	14·0°	14·2°	14·4°	14·6°	14·8°
20	..	..	..	..	..	..	..	..	..	..	..	..	..	..	..
22	..	..	..	..	..	..	..	..	..	..	..	..	..	..	..
24	..	..	..	..	..	..	..	..	..	..	..	..	..	..	..
26	..	..	..	..	..	..	..	..	..	..	..	..	..	..	..
28	5	5	5	5	5	4	..	..	..	..	..	..	..	..	..
30	10	10	9	9	9	8	8	7	7	7	6	6	6	6	5
32	16	16	15	15	15	14	14	13	13	13	12	12	11	11	11
34	23	23	22	21	21	20	20	19	18	18	17	17	16	16	15
36	27	27	26	25	25	24	24	23	23	23	22	22	21	21	20
38	29	29	28	27	27	26	26	25	25	24	24	23	23	22	22
40	31	31	30	29	29	28	28	27	27	26	25	25	25	24	24
42	32	31	31	30	30	29	29	28	28	27	26	26	26	25	25
44	34	33	33	32	32	31	31	30	30	29	28	28	27	26	26
46	35	34	34	33	33	32	32	31	31	30	29	29	28	27	27
48	36	35	35	34	34	33	33	32	32	31	30	30	29	29	29
50	38	37	36	35	35	34	34	33	33	32	32	31	31	30	30
52	39	38	38	37	37	36	36	35	35	34	33	32	32	31	31
54	40	39	39	38	38	37	37	36	36	35	34	33	33	32	32
56	41	40	40	39	39	38	37	36	36	35	35	34	34	33	33
58	43	42	41	40	40	39	39	38	38	37	36	35	35	34	34
60	43	42	42	41	41	40	39	38	38	37	37	36	36	35	35
62	44	43	43	42	42	41	40	39	39	38	38	37	37	36	36
64	45	44	44	43	43	42	41	40	40	39	39	38	38	37	37
66	46	45	45	44	44	43	42	41	41	40	40	39	39	38	38
68	46	45	45	44	44	43	42	41	41	40	40	39	39	38	38
70	47	46	46	45	45	44	43	42	42	41	41	40	40	39	39
72	48	47	47	46	46	45	44	43	43	42	42	41	41	40	40
74	48	47	47	46	46	45	45	44	44	43	43	42	42	41	41
76	49	48	48	47	47	46	46	45	45	44	43	43	43	42	42
78	50	49	49	48	48	47	47	46	46	45	44	43	43	42	42
80	50	49	49	48	48	47	47	46	46	45	45	44	44	43	43

Humidity.—*Continued.*

THE DRY AND WET BULB THERMOMETERS.

15·0°	15·2°	15·4°	15·6°	15·8°	16·0°	16·2°	16·4°	16·6°	16·8°	17·0°	17·2°	17·4°	17·6°	17·8°	18·0°	Dry Bulb Reading. °
..	..	..	..	..	..	..	..	..	..	..	..	..	..	..	..	20
..	..	..	..	..	..	..	..	..	..	..	..	..	..	..	..	22
..	..	..	..	..	..	..	..	..	..	..	..	..	..	..	..	24
..	..	..	..	..	..	..	..	..	..	..	..	..	..	..	..	26
..	..	..	..	..	..	..	..	..	..	..	..	..	..	..	..	28
5	5	5	5	4	4	..	..	..	..	..	..	..	..	..	..	30
10	10	10	10	9	9	9	8	8	8	7	7	7	7	6	6	32
15	15	14	14	14	13	13	13	12	12	12	11	11	11	10	10	34
20	19	19	18	18	17	17	16	16	15	15	14	14	14	13	13	36
21	21	20	20	20	19	19	18	18	17	17	16	16	16	15	15	38
23	23	22	22	22	21	21	20	20	19	19	18	18	18	17	17	40
24	24	23	23	23	22	22	21	21	20	20	19	19	19	18	18	42
25	25	24	24	24	23	23	22	22	21	21	20	20	20	19	19	44
26	26	25	25	25	24	24	23	23	22	21	21	21	21	20	20	46
28	27	27	26	26	25	24	23	23	22	22	22	21	21	20	20	48
29	28	28	27	27	26	26	25	25	24	24	23	22	22	21	21	50
30	29	29	28	28	27	27	26	26	25	25	24	24	24	23	23	52
31	30	30	29	29	28	28	27	27	26	26	25	25	25	24	24	54
33	32	32	31	31	30	29	29	28	28	27	27	26	26	25	25	56
34	33	33	32	32	31	31	30	30	29	29	28	28	28	27	27	58
35	34	34	33	33	32	32	31	31	30	30	29	29	28	28	27	60
35	35	34	34	33	33	31	32	32	31	31	30	30	29	29	28	62
36	36	35	35	34	34	33	32	32	31	31	30	30	29	29	29	64
37	36	36	35	35	34	34	33	33	32	32	32	31	31	30	30	66
38	37	37	36	36	35	35	34	34	33	33	33	32	32	31	31	68
38	38	37	37	36	36	35	35	35	34	34	33	33	32	32	31	70
39	39	38	38	37	37	36	36	35	35	34	34	34	33	33	32	72
40	40	39	39	38	38	37	37	36	36	35	35	35	34	34	33	74
41	40	40	39	39	38	38	37	37	37	36	36	36	35	35	34	76
41	41	40	40	39	39	39	38	38	37	37	36	36	35	35	34	78
42	41	41	40	40	39	39	38	38	37	37	36	36	35	35	33	80

TABLE SHOWING THE PRESSURE OF SATURATED AQUEOUS VAPOUR IN INCHES OF MERCURY AT LATITUDE 45° FOR EACH DEGREE FAHRENHEIT FROM − 30° TO 119°.

°	Inch.	°	Inch.	°	Inch.	°	Inch.	°	Inches.
−30	0·0099	0	0·0440	30	0·1665	60	0·5192	90	1·4128
−29	0·0105	1	0·0461	31	0·1738	61	0·5379	91	1·4578
−28	0·0111	2	0·0482	32	0·1815	62	0·5572	92	1·5040
−27	0·0117	3	0·0504	33	0·1888	63	0·5771	93	1·5514
−26	0·0123	4	0·0527	34	0·1964	64	0·5976	94	1·6001
−25	0·0130	5	0·0551	35	0·2043	65	0·6187	95	1·6502
−24	0·0137	6	0·0577	36	0·2125	66	0·6405	96	1·7017
−23	0·0144	7	0·0604	37	0·2210	67	0·6630	97	1·7546
−22	0·0152	8	0·0632	38	0·2297	68	0·6862	98	1·8088
−21	0·0160	9	0·0661	39	0·2388	69	0·7101	99	1·8646
−20	0·0168	10	0·0691	40	0·2482	70	0·7347	100	1·922
−19	0·0177	11	0·0723	41	0·2579	71	0·7601	101	1·980
−18	0·0186	12	0·0756	42	0·2679	72	0·7862	102	2·041
−17	0·0196	13	0·0790	43	0·2783	73	0·8131	103	2·103
−16	0·0206	14	0·0825	44	0·2890	74	0·8409	104	2·166
−15	0·0217	15	0·0862	45	0·3001	75	0·8695	105	2·231
−14	0·0228	16	0·0901	46	0·3116	76	0·8989	106	2·298
−13	0·0239	17	0·0942	47	0·3235	77	0·9292	107	2·366
−12	0·0251	18	0·0985	48	0·3358	78	0·9604	108	2·437
−11	0·0263	19	0·1030	49	0·3485	79	0·9925	109	2·509
−10	0·0276	20	0·1076	50	0·3616	80	1·0255	110	2·583
− 9	0·0289	21	0·1124	51	0·3751	81	1·0595	111	2·659
− 8	0·0303	22	0·1174	52	0·3891	82	1·0945	112	2·736
− 7	0·0318	23	0·1226	53	0·4036	83	1·1305	113	2·817
− 6	0·0333	24	0·1282	54	0·4186	84	1·1675	114	2·898
− 5	0·0349	25	0·1339	55	0·4341	85	1·2056	115	2·982
− 4	0·0366	26	0·1399	56	0·4501	86	1·2447	116	3·067
− 3	0·0383	27	0·1461	57	0·4666	87	1·2850	117	3·156
− 2	0·0401	28	0·1526	58	0·4836	88	1·3264	118	3·246
− 1	0·0420	29	0·1594	59	0·5011	89	1·3690	119	3·338

GLOBE FOR JANUARY.

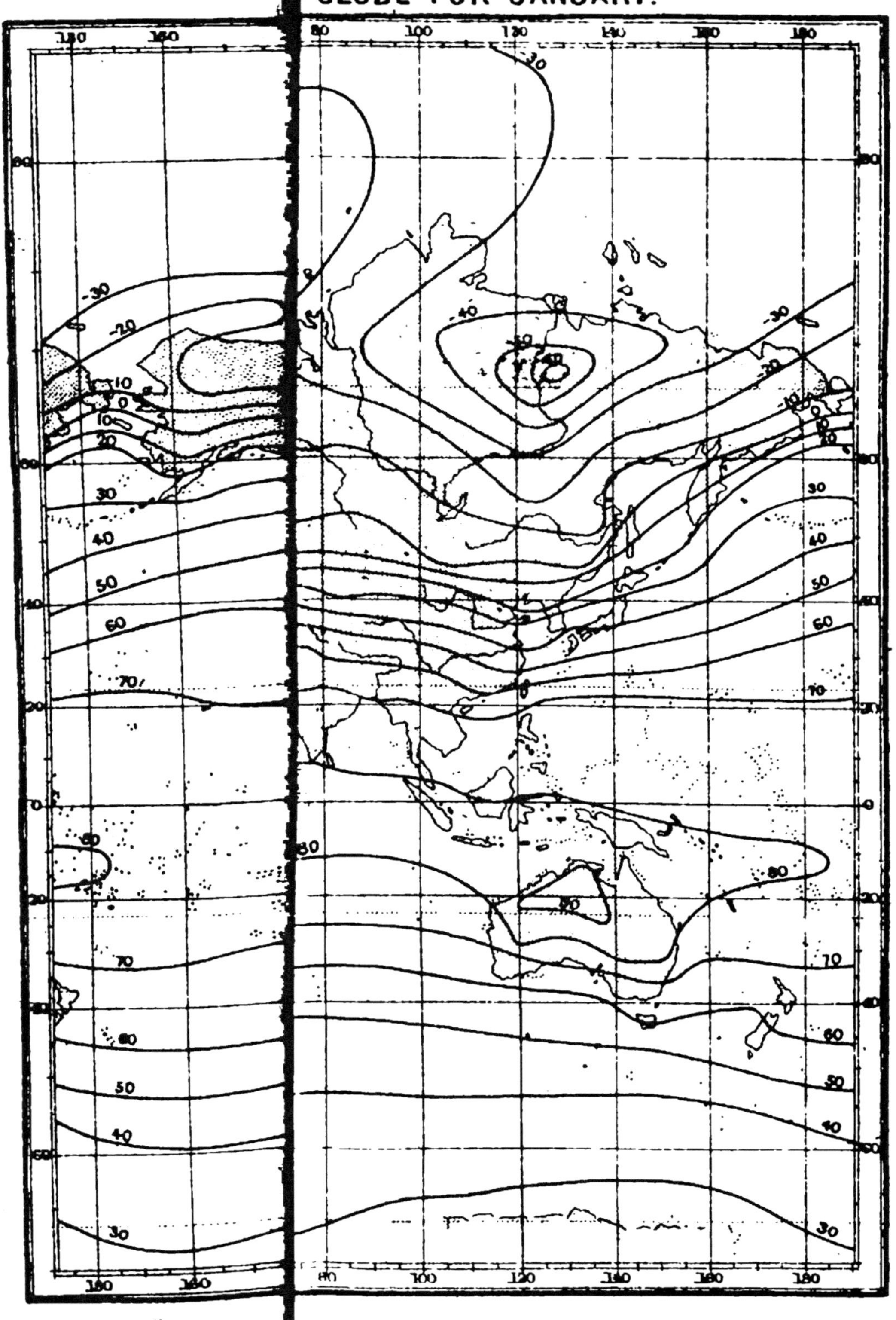

II.

PHOTOGRAPHY.

By J. THOMSON, *Instructor in Photography R.G.S.*

THE photographic camera should form an essential part of the traveller's outfit, as it affords the only trustworthy means of obtaining pictorial records of his journey, and it is also helpful in making the survey of a new region, delineating its contours, its geological and botanical features, and ethnographical types of race. The camera and materials necessary for a journey may be readily obtained, so designed as to minimise space and weight, and in every way so perfectly adapted to the traveller's needs as to ensure successful results in every variety of climate, and render the operation of taking a photograph extremely simple. It is necessary, however, that the traveller should make himself master of the principles involved in the production of a successful photograph, as he will have to depend on his personal effort in exposing and developing the plate, etc. He should also acquire a knowledge of the construction of the camera, to enable him to effect slight repairs when necessary.

In selecting an outfit he must first decide upon the size of plate to be carried, and that need not exceed what is termed "half-plate," $6\frac{1}{2} \times 4\frac{3}{4}$ inches; which is large enough for the best work. The smallest effective size for scientific work may be "quarter-plate," $4\frac{3}{4} \times 3\frac{1}{4}$ inches; in use in many hand-cameras. Negatives on this scale, if perfect, may be enlarged for book illustration, or printed as lantern slides. The two sizes given form a very complete outfit when extra weight may be conveniently carried.

Selecting a Camera.—The larger-sized camera should have a bellows body of Russian leather which folds into small space, its woodwork must be well seasoned to prevent warping, or cracking under a hot sun. The framework should be metal-bound at the corners, and the camera fitted with a rising front and swing back, although the swing back is not indispensable. A reversible back is of advantage, as it enables the operator to take vertical or horizontal views without turning the camera on its side.

The Hand-camera.—Hand-cameras are designed to carry a dozen or more plates or films in flat sheets or in spools, so arranged inside the camera as to be changed after each exposure by simply turning a milled head, or moving a lever. The Key camera is fitted with metal dark slides for carrying plates or films, and is well spoken of by Sir Martin Conway. It may be had either to carry plates or flat films. These metal slides are light, not easily damaged, and offer security against

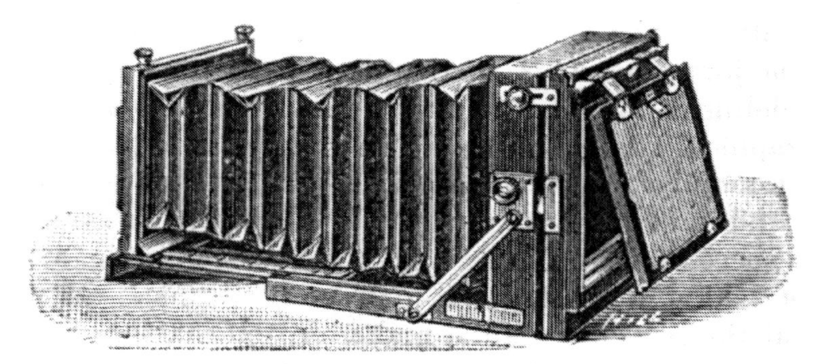

BELLOWS CAMERA.

damp. Rolled films are not recommended for travellers' use in hot and humid climates. Sir Martin Conway says: "A traveller who carries glass plates and flat films will probably bring home a larger percentage of good negatives from a long mountain journey than one who relies upon spools of films." There can be no question about the force and accuracy of this statement. My own experience goes to prove that a camera arranged for glass plates and flat films is best.

The twin-lens hand-camera made by Ross, of New Bond Street, London, offers several advantages in its design. It is fitted with a focal-plain shutter which is in every way simple and effective. The twin lenses are of equal focal length, enabling the object to be photographed to be seen on the same scale as it will appear in the finished negative, so that just what is required may be embraced in the field. It consists of a stout body of thoroughly seasoned hardwood, not easily damaged by rough usage. The principal fittings required for adjusting the instrument are inside, protected by the outer case. The exception to this arrangement is the milled head by means of which the two lenses are focussed at the same time. The lenses are of uniform focal length, so that the image transmitted by the "finder" is a counterpart of the image which falls upon the sensitive plate. The advantage of this is that the object to be taken is seen on the finder-screen to scale exactly as it will appear in the finished photograph. By this means the operator

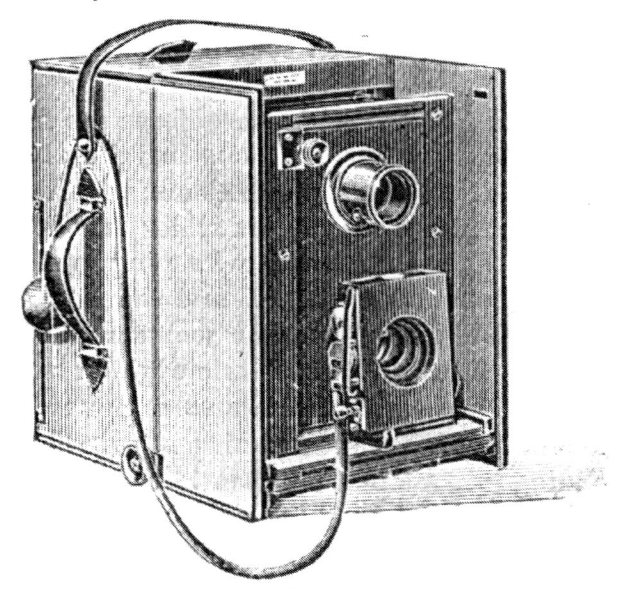

TWIN-LENS CAMERA (OPEN).

has it in his power to place the object in the required position on the screen at the moment of exposure. This is of signal importance if the object is moving about. It must also be noted that this form of camera may be used when the operator is facing at right angles to the object to be photographed. I have frequently found that natives of foreign countries resent the liberty taken of pointing a camera at them, and fly as if they expected to be shot. The slides are each made to hold two plates, or flat films. They are strong, serviceable, and easily managed, while the body of the camera is so arranged as to carry a roll holder,

The camera can also be adapted to stereoscopic work, and fitted for the use of glass plates, flat films, or rolled films.

A light tripod stand should be taken for supporting the camera when longer exposures are required than can be given in the hand. A very satisfactory compromise has been adopted by Sir W. Abney between using the camera in the hand and on a tripod. He rests the camera on top of a walking-stick when making hand exposures, with the result that he overcomes all tremor caused by pulsation, and so secures photographs full of sharp detail.

The plates in the hand-camera may be arranged inside, and exposed one after the other automatically by touching a button, or they may be carried in dark slides.

Slides for Holding the Sensitive Plates.—These are frames which slide into the back of the camera, after the focussing-glass has been removed. They each hold two sensitive plates back to back, with an opaque partition between, so that a dozen plates require six slides or "double backs."

A Focussing-cloth.—This is used for keeping out the light while focussing, being thrown over the camera and the head of the operator. It is generally made of black velvet, but waterproof sheeting is much better. It should have rings sewn on to one edge, or some arrangement by which it may be attached to the camera so as not to be blown away.

Camera-stand.—There are many varieties of tripod stands, with legs either folding or sliding into a small compass. For mountainous country it is of great advantage to have a stand with telescopic legs, as they can be readily altered in length so as to stand firmly on slopes or rocky ground. The smallest size, weighing about 3 lbs., and measuring 33 in. long when closed, and standing about 4 ft. 6 in. high, is steady enough to support a $6\frac{1}{2} \times 4\frac{3}{4}$ camera without perceptible vibration in a moderate wind.

A Small Circular Cup Level, let into the wood of the camera, for levelling the camera on the tripod, is a useful addition.

Lenses.—There are many lenses in the market, and as it is impossible to do good work with an inferior lens, it is necessary to exercise great care in selecting this part of a photographic outfit. Lenses known as rectilinear or symmetrical are useful to a scientific explorer, and are well fitted for producing pictorial effect in his work.

Ordinary portrait lenses are designed specially for rapid work, and this is attained at the cost of qualities in a lens most useful to an explorer. The so-called portrait combination should therefore be avoided.

Rectilinear and symmetrical lenses give true images of objects to be photographed free from distortion, so that straight lines are reproduced as straight lines. In this way they are invaluable where accurate measurements have to be taken from photographs produced by them.

Ross's homocentric lens is one of the latest and most useful lenses. It has a flat field, is free from what is called "coma" and astigmatism, and is so perfectly corrected as to fit it for interior and exterior work alike. It is also a rapid hand-camera lens. The homocentric are made in series to suit all cameras. Other lenses may also be noted, viz.: *c*. Dallmeyer's rapid rectilinear, including about 37°. *d*. Zeiss's anastigmat, made by

HOMOCENTRIC LENS.

Ross, consists of a double front lens and a triple back lens. It is intended for portraits, groups, copying, and general outdoor work. The combinations being brought closely together, gives them great illuminating power. They have an angular aperture of from 858 to 908, and can therefore be used as wide-angle lenses when desired. In consequence of the peculiar system of correction for oblique pencils adopted in these lenses, they behave somewhat differently from the usual types with regard to the mode of compensating the effect of the resulting aberrations between centre and margin of the field. This is, of course, only possible in the case of perfectly plane objects. In all other cases—landscape, instantaneous work, or interiors—the centre should be focussed rather than objects at a distance or foreground.

Focus.—In place of giving a strictly scientific definition of the term focus or "focal length" applied to a lens, it will be sufficient for the scope of this paper to say that focal length means the distance between the diaphragm of a rectilinear or symmetrical lens, and the ground-glass screen of a camera. That is when the image of an object, say one hundred yards in front of the lens, is seen most distinctly on the focussing-screen of the camera.

Exposure Tables.—Exposure tables are based on the focal length of a lens, in relation to the diameter of the diaphragm of a lens. Thus, if the focus is eight inches and diameter of diaphragm one inch, the relationship will be expressed by $\frac{f}{8}$ or by the uniform standard number 4, and so on, as in table.

U. S. Nos.	4	8	16	32	64	128	256
Ratio of Stops	$\frac{f}{8}$	$\frac{f}{11 \cdot 3}$	$\frac{f}{16}$	$\frac{f}{22 \cdot 6}$	$\frac{f}{32}$	$\frac{f}{45 \cdot 2}$	$\frac{f}{64}$

Such tables are useful guides to the relative duration of exposure with diaphragms of different sizes applied to the same lens. They afford no clue, however, to time of exposure to be given with any particular lens or diaphragm. This can be best ascertained by experience, as duration of exposure of a plate or film in the camera depends on the sensitiveness of the plate, the time of day, the sun, the state of the atmosphere, the nearness or distance of the object to be photographed, etc. To take an extreme case of the difference of time required to impress the plate with the image of an exterior view and that of an interior, a landscape open and well lighted may be taken in the fraction of a second, while a dimly-lighted interior with the same lens would require an hour, both being taken with plates coated with the same emulsion. The duration of exposure may be approximately estimated by using an exposure meter such as may be obtained from any photographic dealer's. It is useful to keep an exposure record; a handy book for this purpose is published by Messrs. Burroughs & Welcome. In this book rules are set down for exposure during different months of the year and for different latitudes. These are apt to prove misleading to the amateur. The simplest method of measuring the actinic power of light in any latitude, and at any moment, is by actinometer, giving plate-speeds, focus of lens, etc.

Sensitive Plates or Films.—Gelatine plates are now made commercially by a large number of firms and of great excellence; they keep indefinitely before exposure, and for a long time afterwards and before development and under some circumstances (as for instantaneous pictures, portraits, and dimly-lighted interiors) will give results which could hardly be obtained at all on collodion plates. Gelatine plates are made of various degrees of sensitiveness; the slowest are best for ordinary landscape work. They are generally supplied in parcels of a dozen each, packed face to face with strips of folded paper between opposite edges. The card boxes in which they are usually packed are an insufficient protection against injury and damp. In all cases it is advisable, and for sea voyages and damp climates essential, to have each package of a dozen plates soldered down in a tin case, and afterwards packed in a light wooden box with tow or cotton wool, and the box screwed (not nailed) down. In packing them up again after exposure or after development, a good plan (due to Sir W. Abney) is to provide oneself with a number of cardboard frames exactly the size of the plates, made of strips of card about ¼ in. wide, one of which is inserted between every two plates film to film. The packages thus made up should be soldered down again, and treated with at least as much care as the original plates in packing. Should there be no available means of resoldering the boxes, it will be better to have tin boxes with the lid turned well down, the joinings to be closed by strong well-gummed paper. It will also be well to be provided with a supply of waterproof paper, or cloth, as an additional precaution in packing and in case of emergencies.

Sensitive films in rolls or spools are made by the Eastman and other companies, and may be used successfully in their proper roll-holders when they can be kept perfectly dry in temperate climates. Flat films made by Fitch, Edwards and others have many advantages for travellers. The celluloid of which they are made is very much lighter than glass, and in exposure and development may be treated in the same way as a glass plate. When plates can be carried, the extra weight is compensated for by greater certainty of success, and general excellence in the photographs. The latest novelty has just been put on the market, the Secco Film. It is described as a preparation of the surface of paper in such a manner that the sensitive emulsion adheres closely to it during exposure, so that the finished negative can be readily stripped at any time after drying. The great advantage claimed is that the film has no chemical effect on the emulsion, and is in no way more liable to failure than when the glass plate is used.

How to Keep Plates and Films Dry.—When the traveller has a long journey before him, and the prospect of storing his plates and films for months both before and after exposure, it is of the greatest importance that precautions should be taken against the inroads of damp. This applies with full force when the country to be explored has a hot, humid climate. Plates and films that have absorbed moisture, causing decomposition in the sensitive gelatine coating, are frequently brought back to this country to be developed, and are the most fruitful cause of failure. The remedy is simple, but can only be applied when packing and repacking the plates. Some guarantee should be sought from makers of plates and films that they are packed perfectly dry, and that the packing used is also dry. Assuming that work has to be done in a damp climate and that the plates have been exposed in the camera and require to be repacked, they should be dried in a box containing a small quantity of chloride of calcium. The box used for drying may be also designed to carry the camera and outfit. It should have a lid with a rim of rubber padding, so that by putting the lid on and a weight on it, the box would be fairly air-tight. Stack the exposed plates, or films, in the bottom of box, so separated as to permit the passage of air between. Place a cup or saucer on the bottom of box containing chloride of calcium. (The chloride should be first dried on a piece of iron over a fire.) Put on the lid and allow the plates to remain for an hour or more. Dry all the packing materials, remove the plates from the box and repack. The chloride will have absorbed the moisture in the plates, and rendered them quite dry and safe for preserving for an indefinite length of time.

Apparatus and chemicals for development.—The development of the plates or films after exposure in the camera requires practice and experience in order to secure the best results. Instructions for development are sent out with all commercial plates or papers, but many failures would certainly result from attempting to work by these without some preliminary practice at home. As plates, &c., will keep after exposure (if well protected from damp) for 18 months, or longer if properly packed, it is not, of course, necessary to develop them *en route*, although if the traveller possess sufficient skill, and if ample water-supply and other facilities can be secured, it will be advantageous for many reasons to do so. On a long journey, when the temperature is not too high, use of convenient resting-places may be made to develop from time to time a few plates selected from the whole, both as tests for exposure and as proof that all the apparatus is in order. The following list comprises all that is absolutely required for developing 8 or 10 dozen gelatine plates:— Three papier-mâché dishes, two 3-ounce glass measures, three 6-ounce bottles, containing strong solutions of pyrogallic acid, potassium bromide, and ammonia respectively, 1 lb. hyposulphite of soda, and ¼ lb. alum, both in crystals, 4 or 5 feet of indiarubber tubing and a spring clip, to make a syphon for a water supply from a jug or can, a basin or tub to serve as a sink, a folding rack for draining the plates.

There are several convenient new developing agents in the market: Eikonogen, for example, sold in tubes, and may be used as follows:— Break the tube over a sheet of paper, empty completely both halves by means of pressing the tube between the fingers, withdraw the small pieces of wadding falling therefrom and put the whole quantity of the powder in a bottle containing 100 cubic-centimetres (3½ ounces) distilled water. (Rain-water or soft pump-water may also be used.) After being shaken from 3 to 5 minutes, the powder will dissolve, and the developer is then ready for use. If the plates are over-exposed, increase the quantity of water from 150 to 200 cubic-centimetres, and, if necessary, add a few drops of a solution of bromide of potassium (1 : 10). The developer may be used several times. There are also a number of other developers made up in cartridge form, each cartridge containing all the ingredients necessary for the process of development.

The traveller is recommended for advanced study of photography, such works as that by Sir W. Abney or by W. K. Burton, which may be had from any photographic dealer's.

The aim of the traveller-photographer should be the production of good *negatives*. It often requires years of study on the part of professional operators (with advantages impossible to the traveller) before thoroughly good negatives are habitually produced; and it must not be supposed that a person taking up photography for the first time, in a few hurried moments before departure on a journey, will attain other than very unsatisfactory results.

The operations necessary for taking a picture are briefly as follows:—

Having selected the position from which the view is to be taken (for valuable hints as to the *artistic* production of pictures see Robinson's 'Pictorial Effect in Photography'), the tripod stand is first set up, and the head approximately levelled by means of the pocket level, altering the position or length of the legs as may be necessary. The camera is next screwed on to the stand, and the lens selected which on trial is found to include the required amount of subject. For groups or portraits a long focus lens with wide aperture, such as Dallmeyer's "Rapid rectilinear," 11 in. focus, should be used. The next operation is to focus the picture accurately on the ground-glass screen of the camera. The focussing-cloth is thrown over the head and the camera, so as to exclude the light as much as possible, and while looking at the inverted image on the ground glass, the milled head of the rack adjustment is turned till the image appears as sharp as possible. The camera is now turned about on its vertical axis till it exactly includes the view intended to be taken, and the screw is tightened. It may be necessary to raise or lower the front of the camera carrying the lens in order to include objects at a high or low elevation; if the vertical range of this sliding front is insufficient, the camera must be tilted; but, if this is done, care must be taken to set the focussing-screen vertical again by means of the swing back, and to readjust the focus. The full aperture of the lens should always be used for focussing, and if the image is not sharp all over the plate it will be necessary to insert a diaphragm in the lens, using the largest that will effect the required object. Having then put the cap on the lens, the hinged frame carrying the focussing-glass is turned over, and one of the slides carrying the sensitive plates is inserted in its place. The slides should be exposed as little as possible to the light, especially avoiding direct sunlight; however carefully constructed, it is difficult to make them absolutely light-tight. The shutter of the slide is then withdrawn, and the exposure made by removing the cap from the lens for time exposure, and by a spring shutter for instantaneous work. The time of exposure must be estimated according to circumstances, and it requires considerable experience to judge of it accurately. A record should be kept in a note-book of every plate exposed, giving the number, date, time, exposure, subject, &c. If the plates cannot be developed the same evening, and the slides are wanted for fresh plates, they must be packed up again, and should be numbered. This is best done by marking the number on the back with a bit of dry soap, or in the film with a lead pencil. The image on the plate after exposure is latent and invisible, and has to be developed. This is effected by pouring on the plate, laid in one of the flat dishes, a dilute solution containing pyrogallic acid, ammonia, and potassium bromide. The excellence of the result largely depends on the due proportion between these constituents, and here more experience is perhaps necessary than in any other part of the process. The image having been fully developed, the plate is well washed, and then immersed in a solution of alum, which hardens the film. After another thorough washing it is "fixed" by immersion in a solution of sodium hyposulphite, which dissolves out the unchanged bromide of silver, and, being once more well washed, it is finished, and must be set up in the rack to dry spontaneously. On no account must heat be applied, not even the warmth of sunlight, or the wet film will melt. When dry it must be varnished to protect the film. The printing operations are best deferred till the return home, as they would involve the carriage of a large amount of extra apparatus. It is generally best to get the printing done by a professional printer; but if the traveller prefers to print from his own negatives he will find full instructions in 'The Art and Practice of Silver Printing,' by Robinson and Sir W. Abney.

As regards the expense of a photographic outfit, such as that described above, the following may be taken as average prices for the largest size recommended, namely, for plates 6½ × 4¾ inches:—

Camera, 6 to 8 guineas.

Double slides, about one guinea each.

Lenses, as described above, No. 1, Ross new lens, replacing the Rapid Symmetrical series, £3 5*s.*; No. 2, £3 10*s.*; No. 3, £3 15*s.*

The above may be arranged to pack into a solid leather case, conveniently in the form of a knapsack, measuring about 16 in. wide, 12 in. high, and 5 in. deep. This can easily be carried on the back of one man, and is of a more convenient shape than the cases generally sold for the purpose.

Tripod stand, 25*s.*

Lantern, from 2*s.* to 10*s.*

Gelatine plates, about 3*s.* per dozen.

Apparatus and chemicals for development, about 15*s.*

Total, exclusive of the plates, about £25.

The plates and other apparatus, with the exception of the knapsack and its contents, and the tripod stand, are best packed for travelling in a strong basket, which is much better than a box, being more elastic and lighter. It will weigh, when packed with the apparatus, and a gross of 7½ × 5 plates, about 60 lbs.

Travellers interested in anthropology might read Mr. E. F. im Thurn's paper on the Anthropological Uses of the Camera, published in the 'Proceedings of the Anthropological Society'; also the paper read in May 1900 by Professor Haddon.

Photography in Natural Colours.

It is now possible for the traveller to bring home records of what he has seen in natural colours. There are two or three known methods by which this may be done, but only one available for the work of exploration. The method has been patented by Mr. Ives, and is named Krōmskōp photography. It is briefly as follows. It is based upon the principle that the varied hues in nature are physiologically equivalent to mixtures of three spectrum colours, red, green, and blue-violet. The Krōmskōp photograph is made up of three images which have the appearance of ordinary uncoloured lantern slides, but which are simply three registers, each of the distribution and proportion of a primary colour in the object photographed. Each photograph is thus a colour record—one blue record, another red, and a third green. In order to obtain three colour records, the photographs are taken each through a colour screen, and the result is that each of the primary colours registers only what is required for its own reproduction. Mr. Ives has designed a Krōmskōp for combining the three primary colour records in perfect register, and the result of the combination is a perfect representation of the object photographed in all its natural colours.

Photographs so taken are available for the Lecture-Room, and may be exhibited to an audience on a screen by the use of a properly appointed Triunal Optical Lantern. The first explorer to make use of this most beautiful and perfect of all photographic processes was Mr. Mackinder in his exploration of Mount Kenia. As the result of his successful endeavour, two illustrations in natural colours were published in 'The Geographical Journal,' April 1900.

III.

GEOLOGY.

By THE LATE W. T. BLANFORD, F.R.S.

Revised by PROF. E. J. GARWOOD, F.G.S.

A TRAVELLER who has not devoted some time to studying geology in the field must not be surprised or disappointed if the rocks of any country which he may happen to traverse appear to him a hopeless puzzle. If he desires to investigate the geological structure of an unknown region, he should previously devote some time to mastering, with the aid of a good geological map and description, the details of a well-known tract.

Under the term "Geological Observations," two very distinct types of inquiry are commonly confounded. The first of these, to which the name of Geological Investigation ought properly to be restricted, consists in an examination of the rocks of a country as a whole, so as to enable a geological map, or, at all events, geological sections, to be constructed. This demands a knowledge of rocks (petrology), some acquaintance with the details of geological surveying, and, usually, with the elements of palæontology—a science that, in its turn, requires a preliminary study of biology, and especially of zoology. Despite all these hard terms, any intending traveller who has a taste for geology—if he has none he had better not waste time upon the subject—will find that a few months' study in any good museum, a course of geological lectures, and, above all, a few days in the field with a good geologist, will start him very fairly equipped with the great requisite to all successful scientific investigation, a knowledge of how to observe, and what to observe.

The term "Geological Observations" is, however, often, but incorrectly, used in a second sense, which implies a restriction of the observations to the useful minerals found in any country, or to what is termed economic geology. Here also a preliminary knowledge of the elements of geological science will be found very useful, and will frequently enable the traveller to form much more trustworthy conclusions as to the nature and value of mineral deposits than he could without such a guide. But the essential point is to recognise a valuable mineral when seen, and for this some knowledge of mineralogy is requisite.

Outfit.—The essential articles of a geologist's outfit are neither numerous nor cumbrous. A very large proportion of the known geology of the world has been made out with no more elaborate appliances than a hammer, a pocket compass with a small index to serve as a clinometer, a pocket-lens, a note-book and a pencil. No scientific observer has to depend more on his own knowledge and faculty for observation, and less on instrumental appliances, than a geologist.

The best hammer for general purposes should weigh from 12 to 24 oz. and should have a square flat end, and a straight cutting end—the latter may be horizontal or vertical, according to fancy. The ends should be of steel, not too highly tempered. The hole for the handle should be as large as possible (with a small hole the handles are so weak as to be liable to break), and the handle should be secured in the hole by a wooden wedge, and an iron one driven into and across the wooden one. It is advisable to take a few spare ash handles and iron wedges. Cut a foot-measure in notches on the handle—this is very useful for measuring thickness of beds, &c. It is as well to have more than one hammer in case of loss, and if fossil-collecting is anticipated, at least one heavy hammer, with one end fashioned to serve as a pick, three or four cold chisels of various sizes, and a short crow-bar will be found useful. Excellent geological hammers are those used by the Irish Geological Survey, and made by Kennan & Son, of Dublin. In London, hammers, chisels, &c., may be procured of J. R. Gregory, 88, Charlotte Street, Fitzroy Square; or of Messrs. Buck, 242, Tottenham Court Road.

A very good pocket compass, the shape and size of a watch, with a clinometer arm, is made by Troughton and Simms, 138, Fleet Street. The use of the clinometer is for measuring the angle of dip in rocks. If more accuracy of measurement is required than is afforded by looking at a bed, a section, or a hill-side, and holding the straight-edge attached to the compass parallel to the dip, and if a surface can be found that affords the exact inclination, it is usually practicable, by means of a note-book laid on the rock surface, or, better still, a folding two-foot rule with a slot for sliding in the compass-clinometer, to obtain a plane sufficiently close to that at which the beds dip to enable the angle to be determined with a very short straight-edge. As a rule, except with very low angles of dip, the variation in the inclination of the rocks themselves exceeds the limits of error of the instrument. A little care, however, is necessary in taking dips; for the apparent dip seen in a section, such as is often exposed in a cliff, may differ widely from the true dip, which will only be shown *if the section runs at right angles to the strike of the beds*. Dips seen on the sides of hills at a distance are but rarely correct for the same reason.

A prismatic compass and an aneroid are frequently of great service: the former to determine the position on the map, if one exists, and to aid in making a rough map, if there is none; and the latter to estimate roughly the heights on the road travelled, especially in mountainous countries, and also to measure the thickness of horizontal beds. Both form a part of the outfit of most modern travellers. A good aneroid gives sufficiently accurate determinations of height for a rough but adequate geological section across any country, if the distances are known. The Watkin mountain aneroid of Hicks and Co., Hatton Garden, is the most accurate for considerable heights.

Collections.—Geological specimens require little more than paper and boxes, or biscuit tins, for packing. Occasionally fossils or minerals are fragile, and need tow or grass to protect them from injury; but there is no risk from the animal and vegetable enemies of zoological or botanical collections. The only important point to be borne in mind is that *every specimen should be labelled on the spot*, or, at all events, in the course of the day on which it is collected. Strong paper is best for labels, and these should not be put up in contact with the rock-fragments themselves, or they will be worn by sharp edges and become illegible, if not rubbed to fragments. Always wrap each specimen in paper, or some substitute, then add the label, and then an outer covering, The label,*

* Travellers in tropical countries will do wisely to poison all their labels before using them, to preserve them from attacks of insects and mites. Washing with a very weak solution of corrosive sublimate is an efficient plan. A large number of labels, with the collector's name printed on them, may be taken, and if made of strong thin paper they will not occupy much space. Bank-note paper is well adapted for the purpose. Any writing should be, if possible, in ink; if not, a very hard black pencil should be used. Some tie labels should be added, as paper coverings soon run short.

if nothing else is written, should always record the locality distinctly written.

A collection of rock specimens may show what kinds of rock occur in a country, but the information afforded is very meagre, and, in general, of very small value. Such collections, indeed, unless made by a geologist, and accompanied by notes, are scarcely worth the carriage. If such specimens are taken, care should be used to select them from the rocks in place, *not from loose blocks* that may have been transported from a distance. In certain cases, however, where the traveller does not intend to penetrate beyond the low ground, pebbles brought down by streams give some indication of the strata which occur higher in the drainage basin, and the information may be useful to future explorers, but they should always be labelled as such. No fragments of spar or crystals should be collected merely because they are pretty.

In taking specimens of useful minerals, such as coal or metallic ores, the traveller should always endeavour to procure them himself from the place of occurrence, and if such are brought to him by natives, he should, if practicable, visit the locality whence the samples were procured. The value of all useful minerals depends both on quality and quantity; the former can to some extent be ascertained from a sample, but the amount available can only be estimated after a visit to the locality. Most metallic ores occur in veins or lodes. These were originally cracks in the rock, and have been irregularly filled with minerals, different from those in the neighbourhood. It is, however, very difficult, and often impossible, to estimate from surface examination whether the quantity of ore occurring in veins is likely to prove large; some idea may possibly be obtained if underground workings exist. Many of the ores of iron and manganese, some of those of other metals, and all coal and salt occur in beds, and here it is important to see what is the thickness, and to ascertain whether the mineral is equally pure throughout. Iron ores occur in most countries, and unless very pure and within easy reach of water-carriage, are not likely to be worth transport. The value of salt also depends on facilities for carriage. Coal, however, may be of value anywhere; but it is improbable that seams of less thickness than four or five feet can be of much use, except in countries where there is a skilled mining population and a considerable demand for the mineral. It does not follow because much thinner seams are sufficiently valuable to be worked in Western Europe, that they would pay for extraction in a country where the mechanical arts are less advanced. Still the occurrence of thin seams is worthy of record, as thicker deposits may exist in the neighbourhood. It must not be inferred, however, that a seam of small thickness at the surface will become thicker below. The reverse is equally probable.

A blow-pipe is extremely useful for ascertaining the nature of ores, and for determining minerals generally, and a small blow-pipe case might be added to a traveller's kit, if he thinks it probable that he may meet with minerals in any quantity. But in general they are not to be found in such profusion as to render it difficult to carry away specimens sufficient for determination at leisure. A blow-pipe, too, is of no use to any one unacquainted with the method of employing it, though this is easy to acquire.*

To form a rough idea of the value of iron ore, see whether it is heavy; to form some notion of the quality of coal, pile up a heap and set fire to it. If it does not burn freely, the prospects of the coal being useful are small. It may be anthracitic, and very valuable with proper appliances; but anthracite is not of the same general utility as bituminous coal. Good coal should burn freely, with more or less flame, and should leave but little ash, and it is preferable that the ash should be white, not red, as the latter colour is often due to the presence of pyrites, a deleterious ingredient.

Gold and gems have, as is well known, been procured in considerable quantities from the sands of rivers and alluvial deposits. The deposits known to the natives of any country are often of small value, and the rude methods of washing prevalent in so many lands suffice to afford a fair idea of the wealth or poverty of the sand washed. Gold and, wherever it is found, platinum occur in grains and nuggets, easily recognised by their colour and their being malleable; but gems, such as diamond, ruby, sapphire, are not so easy to tell from less valuable

---

* There are plenty of good works on the use of the blow-pipe. The best are by Plattner and Scheerer, of both of which English translations have been published. Of Von Kobell's tables for the determination of minerals, several translations have appeared. Brush and Penfield 1900 edition is a portable and practical work. A very portable outfit may be obtained from J. T. Letcher, Truro, Cornwall.

minerals. They may be recognised by their crystalline form and hardness. A diamond is usually found in some modification of an octahedron, and the crystalline facets are often curved; rubies and sapphires are really differently coloured varieties of corundum or emery, and occur, when crystalline, in six-sided pyramids or some modification. A diamond is the hardest of known substances; nothing will scratch it, and it will scratch all other minerals. Sapphire will scratch everything except diamond. Topaz will scratch quartz.

In collecting fossils, it is useless to take many specimens of one kind unless carriage is exceptionally plentiful. Two or three good examples of each kind are usually sufficient, but as many kinds as possible should be collected. Great care is necessary that all specimens from one bed be kept distinct from those from another stratum, even if the bed be thin and the fossils in the two beds chiefly the same species. If there is a series of beds, one above the other, all containing fossils, measure the thickness roughly, draw a sketch-section in your note-book, apply a letter or a number to each bed in succession on the sketch, and label the fossils from that bed with the same number or letter.

Remains of Vertebrata, especially of mammals, birds and reptiles, are of great interest; but it is useless to collect fragments of bones without terminations. Skulls are much more important than other bones, and even single teeth are well worth collecting. After skulls, vertebræ are the most useful parts of the skeleton, then limb bones. If complete skeletons are found, they are usually well worth some trouble in transporting. If fossil bones are found abundantly in any locality, and the traveller has no sufficient means of transport, he will do well to carry away a few skulls, or even teeth, and carefully note the locality for the benefit of future geologists and explorers. The soil of limestone caverns, and especially the more or less consolidated loam, rubble, clay, or sand beneath the flooring of stalagmite, if it can be examined, should always be searched for bones, and also for indications of man or his works.

The foregoing remarks are intended for all travellers, especially for those who have paid little or no attention to geology. It would be far beyond the object of the present notes to attempt to give instruction in the methods of geological observation; all who wish to know more fully what questions are especially worthy of attention, should consult the article on Geology by the late Dr. Charles Darwin and Professor J. Phillips in the 'Admiralty Manual of Scientific Enquiry.' But a few hints may be usefully added here for those who have already some knowledge of geology, who do not require to have such terms as dip, strike, fault, or denudation explained to them, and who are sufficiently conversant with geological phenomena to be able to distinguish sedimentary from volcanic, and metamorphic from unaltered rocks, and to recognise granite, gneiss, schist, basalt, trachyte, slate, limestone, sandstone, shale, &c., in the field. Assuming then that a traveller with some knowledge of field geology is making a journey through a tract of the earth's surface, the geology of which is unknown, what will be the best method of procedure and the principal points to which he should direct his attention?

On the whole, the most useful record of a journey, whether intended for publication or merely as a memorandum, is a sketch geological map of the route followed, with the dips and strikes of the rocks and approximate boundaries to the formations, supplemented by notes and sketch-sections. Where, as is commonly the case in mountain-chains, and frequently in less elevated portions of the country, the rocks are much disturbed, and especially if the number of systems exposed is large and the changes frequent, no traveller can expect to do more than gain a very rough and general idea of the succession of beds in detail, and of the structure; but by making excursions in various directions, whenever a halt is practicable, by searching for fossils as a guide to the age and for the identification of beds with each other, and by carefully noting the general dip and strike of the more conspicuous beds, it is often possible, especially if an opportunity occurs of retracing the road followed, or of traversing a parallel route, to make out the structure of a country that at first appears hopelessly intricate. Dense forest is perhaps the worst obstacle to geological exploration; snow is another, though not quite so serious a disadvantage. It is always a good plan to climb commanding peaks; the general direction of beds, obscure from the lower ground, not unfrequently becomes much clearer when they are seen from above.

In level and undulating regions, on the other hand, it frequently happens that enormous tracts of country are occupied by the same formation, and if the rocks are soft, and especially if they are horizontal, or nearly so, little, if any, rock is to be seen in place. In this case water-courses should be searched for sections, and the pebbles found in the stream-beds examined, care being taken not to mistake transported pebbles derived from overlying alluvium or drift for fragments of the underlying rock. Where the same formation prevails over large tracts, it is usually easy, by examining the stones brought down by a stream, to learn whether any other beds occur. It is astonishing how even a small outcrop of hard rock at a remote spot in the area drained by a stream will almost always yield a few fragments that can be detected by walking two or three hundred yards up the stream-bed and carefully examining the pebbles.

Not infrequently different rocks support different vegetation, and by noting the forms that are peculiar, the constitution of hills at a considerable distance may be recognised. Thus some kinds of rock will be found to support evergreen, others deciduous trees, others grass, whilst a fourth kind may be distinguished by the poverty or want of vegetation. It is not well to trust too much to such indications, but they may show which hills require examination and which do not. The form assumed by the outcrop of some hard beds is often characteristic, and may be recognised at a considerable distance.

One most important fact should never be forgotten; mineral character, whether of sedimentary or volcanic rocks, is absolutely worthless as a guide to the age of beds occurring in distant countries. The traveller should never be led to suppose, because a formation, whether sedimentary or volcanic, in a remote part of the world, is mineralogically and structurally identical with another in Europe, or some country of which the geology is well known, that the two are of contemporaneous origin. The blunders that have been made from want of knowledge of this important caution are innumerable.

There are a few points of geological interest well worthy of the investigation of those who traverse unexplored, or partially explored, tracts of the earth's surface. Amongst these are the following:—

Mountain-Chains.—Few, if any, geologists now believe that mountains were simply thrust up from below; all admit that, at least in the majority of cases, where great crumpling of the strata has taken place, there has been lateral movement of the earth's crust. But the causes, extent and date, of the lateral movements are still, to a great degree, matters of conjecture, and every additional series of observations bearing on the question is of importance. There are many mountain-chains of which very little is yet known. In every case good sections are required, drawn as nearly to scale as practicable, through the range from side to side, and including the rocks at each base. The nature and distribution of all volcanic and crystalline rocks, both in the range and throughout the neighbouring areas, are especially noteworthy, and also the relations of the later beds, if any, on the flanks of the mountains, to those constituting the range itself. The derivation of the materials of the former from the latter, and the relative amount of disturbance shown by the two, and by the different members of each, will afford a clue to the date of upheaval; and two or more periods of movement may thus be determined, where intrusive igneous rocks, such as granite, occur, their relations to the surrounding rocks should be carefully noted, and specimens at the contact of the two rocks collected. If altered sedimentary rocks are found these should be traced, if possible, away from the igneous rock until some indication of their age is obtained from included fossils.

The distinction between a contemporaneous lava flow and an intrusive sheet of igneous rock is not always at first sight apparent; if the latter, it may pass from one bed to another or send tongues upwards into the overlying strata. Search should be made in the beds overlying the igneous rock for signs of alteration by heat. Thus limestone may be re-crystallised into marble, or shales altered into flinty hornstone.

Volcanoes and Volcanic Rocks.—It is almost needless to say that any additional information on the distribution of volcanic vents, recent or extinct, is of interest. In the case of extinct vents, the geological date of the last eruptions should be ascertained if practicable. This may sometimes be determined by finding organic remains or sedimentary beds of known age interstratified with the ashes or lava-streams near the base of the volcano.

Coasts.—The subject of the erosion of coasts is now fairly understood, and there is no doubt that the relative importance of this form of denudation was greatly overrated by many geological writers, who took their ideas of geological denudation generally from the phenomena observed in the islands, and on some of the coasts of Western Europe. Still, wherever cliffs occur, they afford good sections, and deserve examination. One question will usually present itself to almost every geological observer, and that is, whether any coast he may be landing upon affords evidence of elevation or depression. In the former case, beds of rolled pebbles or of marine shells, similar to those now living on the shore, may be found at some elevation above high-water mark. Very often the commonest molluscs in raised beds are the kinds occurring in estuaries, which are different from those inhabiting an open coast. Caution is necessary, however, that heaps of shells made by man, or isolated specimens transported by animals (birds or hermit-crabs), or by the wind, be not mistaken for evidence of raised beds.* If the shore is steep, terraces on the hill-sides may mark the levels at which the sea remained in past times, but some care is necessary not to mistake outcrops of hard beds for terraces. If dead shells of species of mollusca, only living in salt-water estuaries, are found in places now beyond the influence of the tide, it is a reasonable inference that elevation has taken place.

The evidence of depression, on the other hand, unless there are buildings or trees partly sunk in the water, is much less readily obtained, and neither trees nor buildings are available as evidence, unless the depression is of comparatively recent date. The best proof is the form of the coast. If deep inlets of moderate breadth occur, with numerous branches, a little examination will frequently show whether such inlets are valleys of subaërial erosion, as they not unfrequently are, that have been depressed below the sea. A good and familiar example of such a depressed valley is to be found at Milford Haven in South Wales. In higher latitudes, the coast should be examined for signs of the action of sea ice, and stones should be collected from icebergs which have drifted from outside the accessible area; the shape of these stones and the proportion of those having only one smoothed side should be noted.

Rivers and River-Plains.—At the present time a question of much interest is the antiquity of existing land-areas, and some light may be thrown upon this, if the relations of existing river-basins to those of past times can be determined. If a stream cuts its way through a high range, it is probable that the stream is of greater antiquity than the range, and either once ran at an elevation higher than the crest of the ridge now traversed, or else has cut its way through the range gradually during the slow elevation of the latter. Where a river traverses a great alluvial plain, it may fairly be inferred that a long time has been occupied in the

* In high latitudes care must be exercised in distinguishing between true raised beaches and ridges of beach material pushed up by the pressure of shore ice in winter.

accumulation of the deposits to form the plain; but it remains to be seen whether those deposits are not partly marine or lacustrine. If upheaval has taken place over any portion of the plain, or if the river has cut its bed deeper, sections may be exposed, and these should always be examined for fossil remains. Bones of extinct animals are not unfrequently found in such deposits.*

Lakes and Tarns.—The mode of origin of lakes is always a subject of considerable geological interest, and any evidence which bears on the origin of a particular lake should be carefully noted. Lakes may be divided broadly into two classes: (1) Rock basins, (2) impounded hollows. Lakes of the latter class may, as a rule, be readily recognised and accounted for. The material forming the barrier may be due to a moraine, screes, or a landslip, or may result from the presence of a glacier in the main valley damming back the drainage of a lateral tributary. When lakes occur near the summit of a pass, they may often be traced to deposition of delta material on the floor of the valley brought down by a tributary stream. In this case, the inability of the stream to remove the material may often be traced to the abstraction of the head waters of the valley by the encroachment of the stream on the other side of the watershed. Rock basins, on the other hand, are frequently difficult to account for. They may occupy volcanic craters, or lie in areas of special depression (earthquake districts), or synclinal folds. They may be due to upheaval of old valley systems, causing reversal of drainage, or to subsidence at the upper end of a valley.

---

* In some valleys the lateral tributaries enter at a higher level than the floor of the main valley, owing to the overdeepening of the latter; this overdeepening has been attributed to four different causes:—(1) Inequality in the hardness of the rocks, especially when the main valley runs along the strike; (2) the presence of joints, or other lines of weakness; (3) increased erosion of the main valley by river action owing to the elevation of the upper end of the district, while the lateral valleys were merely tilted sideways, or were protected by the presence of ice or lakes; (4) erosion of the main valley by ice more rapidly than its tributaries, owing to the ice being thicker in the former. The problem is complicated by the fact that the upper end of the main valley is frequently a 'hanging' valley. Any evidence bearing on the formation of these 'hanging' valleys should be carefully noted.

In special cases they may be due to the solution of soluble rock (rock-salt, gypsum, limestone, or dolomite). Many not otherwise explicable have been attributed to ice erosion during glacial periods, and it is still a moot point how far these lakes are due to partial changes in the elevation of the country, some observers having adopted, while many others dispute, the views of the late Sir A. Ramsay, who believed all these hollows to have been scooped out by ice moving over the surface in the form of a glacier or an ice-sheet. The origin of any lake met with should, if possible, be investigated and assigned to one of these causes.

Evidence of Glacial Action.—Closely connected with the subject of lakes is that of glacial evidence generally. There is probably no geological question which has produced more speculation of late years than the inquiry into the traces of a comparatively recent cold period in the earth's history, and the former occurrence of similar glacial epochs at regular or irregular intervals of geological time.

The evidence of the last glacial epoch may be traced in two ways—by the form of the surface, which has been modified by the action of ice, and by changes that have taken place in the fauna and flora of the country in consequence of the alteration in the climate. The effects of an ice-sheet, like that now occurring in Greenland, if such formerly existed in comparatively low latitudes, must have been to round off, score and polish the rocks of the country in a peculiar manner, easily recognised by those familiar with glaciated areas. Care should be taken that the peculiar scoring and grooving of rock surfaces produced by the action of sand transported by the wind be not mistaken for glacial evidence. Cases also occur where movement among a mass of unconsolidated conglomerate or scree material has produced striation of the pebbles; in this case, however, careful observation will disclose a similar striation in the material of the matrix as well. Glaciers, properly so called, are confined to hilly or mountainous countries, and the valleys formerly occupied by them retain more or less the form of the letter U instead of taking the shape of the letter V, as they do when they have been cut out by running water. The sides of the valley, when modified by a glacier, have a tendency to assume the form of slopes unbroken by ravines, and with all ridges planed away or rounded, whilst in ordinary valleys of erosion by water, the sides consist of a series of side valleys or ravines, divided from each other by sharp ridges running down to the main valley. Large and small masses of rock, preserving to a considerable extent an angular form, but frequently polished and grooved by being ground against the sides or bottom of the valley, are carried down by the ice, and either left behind, perched up high on the slopes of the valley, or accumulated in a vast heap or bank, known as a terminal moraine, at the spot where the ice has terminated, or as lateral moraines on the sides of the valley. The nature of the rock will usually show whether the fragments on the side of a hill or at the bottom of a valley are derived from the higher parts of the drainage area, or whether they have merely fallen down from the neighbouring slopes. In the latter case, they may be due to landslips; in the former, their shape and the erosion they have undergone will aid in showing whether they have been transported by water or ice.

The surfaces that have been modified by earlier glacial epochs must in general have been long since removed by other denuding agencies. The most important evidence of former ice action consists in the occurrence, embedded in fine sediment, of large boulders, occasionally preserving marks of polish and striation, and usually, though not always, angular. Accumulations of this kind afford evidence of transport by two different agencies, water, which has brought the silt, and ice, which has carried the boulders. If the water had been in rapid movement, and thus capable of moving the boulders, it would have carried away fine silt or sand, instead of depositing it. Evidence of ice action has thus been traced equally in the boulder clay of North-Western Europe, and in the Palæozoic boulder beds of India, South Africa, and Australia, and probably of South America.

It is well to search in all mountain ranges for traces of glacial action. In many mountain chains, even in comparatively low latitudes, proofs have been found of the existence of glaciers, at a much lower level than at present, dating from a comparatively recent geological period, whilst in other mountain regions none have been recognised. The question also whether glacial action has been contemporaneous in the two hemispheres is of the greatest importance, and the evidence hitherto adduced is of a very conflicting character.

Deserts.—The great sandy or salt plains, with a more or less barren surface, that occupy a large area in the interior of several continents, have only of late years received due attention from geologists. A great thickness of deposits must occur in many of these vast, nearly level, tracts, for the underlying rocks are often completely concealed over immense areas. The investigation of the deposits is frequently a matter of great difficulty for want of sections; but, where practicable, a careful examination should be made, and exact descriptions of the formations exposed recorded. Some, at all events, of these beds appear to be entirely deposited from the air, and consist of the decomposed surfaces of rocks and the sand and silt from stream deposits, carried up by wind and then redeposited on the surface of the country. Such deposits are very fine, formed of well-rounded grains, and, as a rule, destitute of stratification. The geologist who has especially described these formations, Baron F. von Richthofen, in his work on China, attributes to the loess of the Rhine and Danube valleys a similar origin. It is usual to find beds due to water-action, rain-wash and stream-deposits, interstratified with the subaerial accumulations. Further observations on these formations are desirable. The occurrence of blown sands, the origin of these accumulations, and the peculiar ridges they assume, usually at right angles, but in some remarkable cases parallel to the prevailing winds, are questions deserving of additional elucidation.

Early History of Man in Tropical Climates.—Very little has been discovered as to the races of men formerly inhabiting tropical regions. It is evident that a race unacquainted with fire could only have existed in a country where suitable food was procurable throughout the year, and this must have been in a region possessing a climate like that found in parts of the tropics at the present day. It is possible that an investigation of the cave deposits in the tropics may throw some light on this subject. "Kitchen middens," as they are termed—the mounds that have once been the refuse heaps of human habitations—are also worthy of careful examination.

Permanence of Ocean-Basins.—Within the last few years some geologists have adopted the theory that all the deep-sea area has been the same from the earliest geological times, and that the distinction between the depressions occupied by the oceans and the remaining undepressed portion of the earth's crust, constituting the continents and the shallow seas around their coasts, is permanent. This view is very far from being universally or even generally accepted amongst geologists, although many who hesitate to accept the theory as a whole admit that parts of the oceans may have been depressions since the earth's crust was first consolidated.

The argument on both sides depends upon theories to which travellers can contribute but little except by observations on the geology, fauna, and flora of oceanic islands, and by the investigation of coral-reefs and especially of atolls. In ranges of hills or mountains near the coast both of continents and islands and in all tracts where evidence of recent elevation exists, search should be made for deep-sea deposits. These are fine calcareous or argillaceous beds, often containing small Foraminifera or Radiolaria, which, however, are generally extremely minute, and require microscopical examination for detection. If any beds of consolidated calcareous or siliceous ooze or especially if red or gray clay (in older rocks, slate, or even quartzite) be found associated with pelagic deposits, such as coral limestone, a few small fragments of the beds should always be brought away for examination, and any distinct fossil remains found in such beds, for instance echinoderms (sea-urchins or star-fishes) or sharks' teeth, should be carefully preserved with some of the matrix. Deep-sea deposits have recently been discovered in several parts of the world, for instance, the West Indies, the Solomon Islands, the islands of Torres Straits and Southern Australia, as well as in Europe.

Atolls or Coral-Islands.—Each of the remarkable coral islands of the Pacific and Indian oceans consists usually of an irregular ring, part or the whole of which is a few feet above the sea, and which encircles an inner lagoon of no great depth. The outer margin of the reef around each island slopes rapidly, sometimes precipitately, to a depth of, usually, several hundred fathoms. Darwin, taking these facts into consideration, together with the circumstance that no coral-reefs are known to be formed at a greater depth than about 15 to 20 fathoms (90 to 120 feet), showed that all the facts of the case could be explained by the theory that coral-islands were formed in areas of subsidence. This view was generally accepted until Prof. A. Agassiz, Sir John Murray, and other writers brought forward evidence in favour of coral-islands being founded on shoals that may be areas of elevation.

Much light has been thrown on this subject by recent exploration. Two instances in especial may be mentioned. The examination of Christmas Island in the Indian Ocean, South of Sumatra, by Mr. Andrews, has shown it to be a raised atoll, founded on a volcanic base, whilst, on the other hand, borings on the atoll of Funafuti, one of the Marshall Islands in the Western Pacific, carried to a depth of over 1000 feet on the ring itself, and to 245 feet below sea-level in the middle of the lagoon, have yielded results which, in the opinion of the geologists engaged, Prof. Sollas and Prof. Edgeworth David, completely confirm Darwin's theory.

It is probable that atolls originate in more than one way, some being formed in rising or stationary tracts, others in areas of depression. The important question, from a geographical point of view, is not so much how isolated atolls were formed as whether the great tracts in the Pacific and Indian oceans in which no islands occur except atolls, for instance, the Marshall, Gilbert and Low archipelagoes in the former, and the Laccadives and Maldives in the latter, have been areas of extensive subsidence during the later geological periods. Further evidence on this question may perhaps in time be furnished by additional borings, for one of which an island of the Maldive group would furnish an excellent locality, since there is in this case independent evidence to indicate that the archipelago occupies part of a sunken tract. Meantime any additional details would be useful, such as careful soundings around those atolls which have not been fully surveyed, so as to give an accurate profile of the sea-bottom in the neighbourhood.

MEMORANDUM ON GLACIER OBSERVATIONS.

The recent movements of glaciers may be noted by the following signs:—

When the ice is advancing, the glaciers generally have a more convex outline, the icefalls are more broken into towers and spires, and piles of fresh rubbish are found shot over the grass of the lower moraines. Moraines which have been comparatively recently deposited by advancing ice are disturbed, show cracks, and are obviously being pushed forward or aside by the glacier. There is a tendency for the glacier to terminate in a vertical front, or "Chinese wall," as distinguished from the sloping snout of a stationary or retreating glacier, owing to the more rapid advance of the upper layers. If the advance is rapid the overhanging layers will fall, and an ice talus will collect in front. An inspection of the ice will often disclose horizontal lines of sheering which in a side view are seen to rise towards the terminal wall. Search should be made along these sheer planes for included moraine material, and any proofs of elevation of this material noted.

When the ice is in retreat it terminates in a gently sloping snout. The marks of its further recent extension are seen fringing the glacier both at the end and sides in their lower portions, the glacier fails to fill its former bed, and bare stony tracts, often interspersed with pools or lakelets, lie between the end of the glacier and the mounds of recent terminal moraines.

Where a glacier has retreated to any considerable extent, careful observations of the form of its bed are of value. What is the nature of the rock surfaces exposed—convex or concave; are they rubbed smooth on their leesides; how far have the contours of the cliffs or slopes, or the sides of any gorge, been modified where they have been subjected to ice-friction? Is there any evidence that the ice has flowed over large boulders, or loose soils, such as gravel, without disturbing them? How has it affected rocks of different hardness, for instance, veins of quartz in a less hard rock? Generally, do the appearances indicate that the glacier has excavated, or only abraded and polished its bed; that it has scooped out new rock-basins, or only cleaned out, scratched, and preserved from filling-up by alluvial deposits or earthslips, existing basins? What is the general character of the valley bottom and the slopes above and below the most conspicuous ancient moraines?

The traveller or surveyor should, if possible, paint a mark and date on any conspicuous rock *in situ* parallel with the termination of the glacier at the time of his visit, marking the distance of the ice from it. The next visitor will then be able to measure the movement that has taken place since his predecessor's visit. Leaving out of question elaborate trigonometrical methods, such, for instance, as have been carried out on the Rhone Glacier in Switzerland, the following plan gives very valuable results, and demands no other instruments than a small jar of paint, a brush, a measuring tape, and a pocket compass. To ascertain the recent retreat of a glacier, measure the distance from the end of the ice in front of the longitudinal axis of the glacier to the most advanced terminal moraine, where vegetation first shows itself. The bare ground recently left by glaciers is easily recognisable. The diminution of volume is best measured by ascertaining the height of bare soil left on the sides of the lateral moraines in the portion of the glacier within the zone of vegetation. All photographic representations of the glacier end, and of the ground which has been freed from the glacier ice, are of great value. Those will be of most service that show the position of the glacier-snout with relation to some conspicuous rock or other feature in the local scenery. Each photograph should be dated, and the bearings and distance of the camera with reference to any such feature accurately noted.

It is very important to investigate the state of various glaciers as regards advance or retreat. Neighbouring glaciers often furnish very different results in this respect, owing to the fact that steep glaciers anticipate in their oscillations those the beds of which are less inclined. To ascertain the oscillations of glaciers, it is necessary to fix the actual position of the ice-snout at the end of the glacier with the greatest accuracy. Two methods can be employed for this purpose, either of which may be selected according to circumstances.

Paint some signs on large boulders, not too far from the end of the glacier, and measure their distance from it by a tape (Richter's system), or build a low wall of stones of a few yards in length, and, say 15 to 20 inches in height, some distance from the ice-end, and measure this distance (Gosset's system). It is to be recommended that the stones of these walls should also be painted. If the traveller himself returns after some interval—even after only two or three weeks—he will be able to judge of the movement of the glacier, and he will have laid down a basis for further observations by future travellers.

One of the results most to be desired is an exact knowledge of the dates:

I. Of the maximum extension of the ice.
II. Of the commencement of retreat.
III. Of the minimum.
IV. Of the commencement of fresh increase.

In dealing with a mountain group, therefore, the traveller should note (where he can get the information as to the past) the date of the commencement of the actual movement of *each glacier*, and in all cases whether the ice is in advance, or retreat, or stationary. Of course the rate of forward movement, or velocity of the ice, and the oscillations in the extension of the ice must be kept carefully distinct. Observations should also be made on the presence of blue bands, and their relation to the lines of stratification in the névé noted.

Should time and circumstances permit, a series of observations of the velocity of the ice is of value. These may be made after Tyndall's method, by planting a line of sticks across the glacier, or by painting marks on boulders, the position of which relatively to ascertained points on the mountain-side has been accurately fixed. The size of the glacier that is, the area of its basin and its length, also the slope of its bed above, as well as at the point measured, should be noted. The rate of movement of the ice appears to be connected both with the volume of the glacier and the inclination of its bed, and is considerably affected by temperature. Thus a rise in temperature may even be accompanied by a temporary advance, but in this case a sagging will take place higher in the névé, producing a concave cross-section, showing that the advance is not due to increased snowfall, but to a decrease in the viscosity of the ice.

The advance or retreat of a glacier are not the only factors to take into account in estimating the decrease or diminution of the volume of ice. The highest level of the transverse convexity of the glacier at various points in its course should also be noted. If the glacier is bounded laterally by rocky walls, marks may conveniently be painted on these, opposite one another. The vertical height of the marks above the ice at the sides should be noted, and the date of the observation recorded.

A society, entitled the Commission Internationale des Glaciers, has been formed to promote the study of glacial movements and other points of interest and importance. Their reports are published in the 'Archives des Sciences,' Geneva, the journal of the Swiss Alpine Club, and have been summarised by M. Rabot and translated into English in the 'Geological Journal,' Washington. Both of these are in the library of the Alpine Club.

IV.

NATURAL HISTORY.*

By the late H. W. Bates, F.R.S.

Revised by W. R. Ogilvie-Grant,
British Museum (Natural History).

In the present state of Biological Science, travellers who intend to devote themselves specially to the zoological or botanical investigation of new or little-known countries, require to be trained for the work beforehand, and are consequently well-informed as to methods and appliances. It is not for them that these 'Hints' are drawn up, but for general travellers and explorers, who, whilst engaged chiefly in survey, wish to know how best to profit by their opportunities of benefiting science by collecting examples of new or rare species, and how to preserve and safely transmit their specimens. The observations refer only to explorations by land.

Outfit†—

A magazine sporting rifle, either Mannlicher ·256, or Lee-Enfield ·303, and ammunition (solid and soft-nosed, split or hollow-point bullets).

Double-barrel 12-bore gun—for choice a Keeper's hammerless non-ejector gun and ammunition (Nos. 3 and 5 shot).‡

---

* Travellers who intend to make collections of zoological specimens are advised to call on the Secretary of the British Museum (Natural History), Cromwell Road, S.W. He will be happy to supply any information required.

† General travelling outfit can be procured from Messrs. Silver & Co., 67 Cornhill, E.C.

‡ Ejector guns are apt to get out of order in sandy countries.

Double-barrel ·410 hammer-gun and ammunition (Nos. 8, 10 and 12 shot).

Cartridges with a full charge of powder and No. 8 shot will kill larger birds at a considerable distance; those with Nos. 10 and 12 shot are specially loaded for collecting smaller birds at closer quarters, and may be used at distances of from fifteen to twenty-five yards without damaging the specimens. The loads are: ⅜ oz. No. 12 shot, ¾ dr. black powder, or 9 grains Schultz; ½ oz. No. 10 shot, ¾ dr. black powder, or 9 grains Schultz. In damp climates black powder will be found more reliable. If shot larger than No. 8 is used with this gun, the pattern is erratic and the result uncertain. Messrs. Cogswell and Harrison, 226 Strand, W.C., have had considerable experience in loading cartridges for these guns, and their Nos. 10 and 12 shot is even and carefully screened.

No. 1 saloon pistol and ammunition (turned-over caps with dust shot). This pistol will be found of great value for collecting small birds at very close quarters in thick cover, or the more active lizards. It can be easily carried in a holster on a belt, to which is also attached a small pouch for ammunition. Small brass turned-over caps, loaded with a pinch of dust-shot, can be had of Messrs. Cogswell and Harrison.

Telescope.

Binocular field-glasses (either Zeiss or Goerz).

Traps can be procured from Mr. S. J. Beckett, 39 Dresden Road, Highgate, N. (apply, in the first place, by letter). For catching hyænas, etc., large traps are supplied; for smaller mammals, Brailsford live-traps and Cyclone and Schuyler break-back traps are the best, and should be procured in several sizes. Pitfall-traps made by sinking earthenware jars flush with the ground are also useful.

Several air-tight zinc-lined collecting boxes, fitted with light wooden trays. These are made by Messrs. Lovelace and Co., 20 Gloucester Road, S.W., and will be found extremely serviceable for packing, storing, and transmitting skins of smaller mammals and birds. When starting they can be packed full of stores. Uniform cases may be substituted, but, unless protected by wooden travelling-covers, are liable to get damaged and to admit insects.

A set of carpenter's tools; screws and French nails of various sizes.

A set of soldering irons and soft solder, etc.

*For skinning large Mammals**—

2 shoemakers' knives.†
3 large scalpels.*
1 small saw.†
1 pair cutting-pincers.†
1 pair pliers.†

*For skinning small Mammals and Birds**—

Taxidermist's box containing:

6 scalpels.†
1 oil-stone.†
2 pairs of scissors (one short-bladed, one dissecting).†
2 pairs of tow-pliers (like curling-irons, for inserting tow or cotton-wool into the necks of bird-skins).†
1 pair of forceps.†
1 pair of tweezers.†
1 pair cutting-pliers.†
1 pair compasses.†
1 2-foot rule marked in tenths of an inch.†
1 millimetre rule.
3 darning needles.†
Needles and thread.
1 tin of mixed pins.
Bundles of galvanized wire (for inserting in tails of small mammals and in necks of bird-skins).
3 hog-hair brushes for applying arsenical soap.
2 egg-drills (two sizes).†
2 blow-pipes (two sizes).†
2 pencils.
1 small biscuit tin of fine boxwood sawdust.
1 2-lb. tin of candle naphthaline.
1 2-lb. tin crystal.
1 2-lb. tin plaster of Paris.

8 1-lb. tins of arsenical soap (in treacle tins), or the same weight of powdered arsenic.
1 2-lb. tin burnt alum.
1000 labels for bird-skins.
500 labels for small mammals.
2 bundles of wool.
Tow (which can be used as packing).
6 pieces of fine mesh wire-netting (for making cages, etc.).

* All the articles mentioned in this list can be best procured from Mr. S. J. Beckett, 39 Dresden Road, Highgate, N. Apply, in first place, by letter.
† Those marked with a cross (†) can be obtained from Messrs. Buck, 242 Tottenham Court Road, W.C.

For preserving Reptiles, Batrachians, and Fish—
These should be preserved in spirits whenever practicable.

For larger specimens—
2 or 4 four-sided copper tanks with a round opening of six inches diameter, closed with a screw cover. Each tank is fitted into a strong wooden case with a lid secured by hinges and lock, and furnished on two opposite sides with a rope handle.

For smaller specimens liable to be damaged—
2 or 4 wooden cases divided into four or six compartments, each containing a pickle jar, with glass rubber-edged stopper secured by an adjustable iron fastening.
3 dozen or more corked glass tubes of various sizes.
A supply of sheets of zinc, soldering irons and soft solder (to make extra tanks).
1000 parchment labels bearing numbers.
If the above specially constructed zinc tanks cannot be procured, wide-mouthed earthenware jars, tightly closed with cork or rubber, may be used.
The glass tubes will be found specially useful for collecting Arachnida, Myriopoda, small Mollusca, etc.

For collecting Insects—

Store boxes lined with cork-carpet.

Triangular envelopes of smooth foolscap paper for butterflies and thin-bodied moths.

Card fly-discs.

Entomological pins (boxes of various sizes).

2 zinc oval pocket boxes lined with cork-carpet (two sizes).

6 or more sets (four each) of glass-bottomed pill-boxes covered with jaconet.

2 pairs of entomological forceps.

4 insect killing-bottles (two sizes) in leather cases. (Hinton & Co., 38 Bedford Street, Strand, W.C.).

A small bottle of oxalic acid, with stabbing quill fixed in cork, for killing large insects.

2 flat killing-bottles for beetles, with glass tube passing through the cork and fitted with a cork plug.

2 small entomological lamps.

6 tins of bicycle oil.

1 dozen 1-lb. tins of golden syrup or treacle.

2 sugar brushes.

1 3-oz. bottle of acetate of amyl ("Essence of Pear Drops," for mixing with treacle) (Hinton & Co.).

3 butterfly-nets (two round, one kite-shaped) with extra bags.*

1 water-net (for aquatic insects).

Several pieces of cork-carpet about a foot square (for making extra store boxes).

1 pocket lens.

2 large pickle-jars of carbolized sawdust to be used as packing for beetles after they have been killed.

3 dozen corked glass tubes (three sizes).†

† All the above-mentioned articles for collecting insects may be procured from Mr. Janson, 44 Great Russell Street, W., or from Messrs. Watkins & Doncaster, 36 Strand, W.C., or from Miss E. M. Sharpe, 4 Barrowgate Road, Chiswick.

---

* The shafts of old golf-drivers, when shortened to about 2 feet 6 inches in length, make the most perfect handles for nets.

For instructions regarding the collecting and preservation of specimens in all branches of natural history, travellers and others are recommended to provide themselves with a 'Handbook of Instructions for Collectors,' issued by the British Museum (Natural History). With illustrations. Second edition, 1904. *Price* 1*s.* 6*d.*

In humid tropical countries, where the ubiquitous ants are likely to destroy specimens before they are ready to be packed away, drying-cages, suspended from the roof of a hut or tent, are absolutely necessary. These can be readily made from old packing-cases, but a few square feet of wire gauze must be provided for the back and front of the cages, and the cord by which they are suspended must be threaded through a small calibash containing oil, or, better still, naphthaline, to prevent ants from descending from the roof. The cages may be so arranged as to be taken to pieces and put together again readily; one, for birds, should be about 2 feet 6 inches long by 1 foot 6 inches high and 1 foot broad; the other, for insects and other small specimens, may be about one-third less. They should have folding doors in front, with panels of wire gauze, and the backs wholly of the latter material; the sides fitted with racks to hold six or eight plain shelves, which in the smaller cage should be covered with cork, or any soft wood that can be obtained in tropical countries. A strong ring fixed in the top of the cage, with a cord having a hook attached at the end by which to hang it in an airy place, will keep the contained specimens out of harm's way until they are quite dry, when they may be stowed away in suitable close-fitting boxes. An even simpler and perfectly effective plan is to take a number of pieces of stout wire each about 18 inches long, bend each end into a loop, and round the middle solder a funnel-shaped piece of tin to contain powdered candle-naphthaline. The upper loop of each wire can be secured with string to a rafter, and between the lower loops flat boards, or a series of boards, can be suspended on which skins may be placed to dry. This method has been proved to be safe, no ants ever venturing to cross the naphthaline.

A few yards of india-rubber waterproof sheeting may be found useful as a temporary covering to collections in wet weather or in crossing rivers.

To those who have had little or no experience in field-collecting it may be useful to give some idea of the impedimenta considered absolutely essential for an ordinary day's work when it is desired to collect, as far as possible, in all branches of zoology, and the most convenient means of disposing of such. The collector should be accompanied, if possible, by two intelligent natives to act as bearers, who may, with patience, be developed into useful assistant-collectors.

A 12-bore gun or rifle, according to circumstances, with ammunition, to be carried by the first bearer.

A ·410 collecting gun and twenty-five cartridges, including a few cartridges with full loads of 8 shot: the cartridges to be carried by self, the gun, when not in use, to be carried by the second bearer. The 10 shot and 12 shot ·410 cartridges should be carried in the two lower waistcoat pockets, where they are easily accessible, and the few 8 shot cartridges in some other pocket.

Saloon pistol in holster with pouch for carrying ammunition on belt, strapped round the waist, so that it can be easily and quickly made use of.

A butterfly-net carried by self, or by the second bearer if not in use.

Large oval zinc pocket-box lined with cork carpet, containing a stock of pins, both large and small, stuck in one side of the cork carpet ready for use; to be carried in the right-hand side-pocket of the coat.

Entomological forceps, pointed forceps, and larger forceps for picking up scorpions, large spiders, etc., and entomological killing-bottle; to be carried in the left-hand side-pocket of the coat.

Two larger-sized corked glass tubes half-full of spirits and a camel's-hair brush to be carried in the right and left top waistcoat-pockets, for collecting small spiders, etc. It is difficult to pick up the swift-running ground-haunting species without injury, but, by dipping the brush in spirits and placing it on them, they are at once stupefied, and may then be easily transferred to the tube.

The spirit is also required for saturating the small plug of cotton-wool to be pushed down the throat of each bird as soon as it has been killed; the gape should then be plugged with dry cotton. Specimens thus prepared may be carried, even in a hot climate, for several hours without deteriorating.

A game bag carried by the second bearer, with a large supply of paper for wrapping up birds when shot.

Several sets of glass-top pill-boxes carried in the 'hare'-pocket, or in one of the pockets of the game-bag, to which they can be transferred when filled.

A pickle pot in Willesden canvas or basket-work cover with handle, half filled with spirits, for collecting small snakes, lizards, frogs, scorpions, etc.; to be carried by the first bearer, who should be instructed how to hunt for reptiles, etc.

If possible, a beetle-killing bottle should be added to the above impedimenta, and may be carried in the left-hand breast-pocket of the coat.

A Norfolk coat is a most useful article of clothing, and should be provided with a deep 'hare' pocket running round the skirt and divided in the middle.

After a very short time the collector will be able to find any article he may require by instinct, and without loss of time. To have a pocket for each article, and to know where it is, saves an infinity of trouble.

At daylight the traps should be visited, and any specimens to be preserved should at once be sent back into camp.

Collecting should always, if possible, be vigorously prosecuted during the early morning hours, when birds, etc., are feeding, and are much more easily procured.

When butterflies and thick-bodied moths are placed in the killing-bottle, care should be taken to see that they die with their wings turned the 'right way,' i.e., with the underside outermost. Those which die with the upper side outermost should be at once reversed with the aid of a pin or the sharp-pointed forceps, and then replaced in the bottle. If not attended to at once they become rigid, and the wings get rubbed and spoilt before they are quite dead.

Where and what to collect.—The countries which are now the least known with regard to their natural history are New Guinea and some of the large islands to the east of it, East Sumatra, the highlands of Mindanao and other Philippine Islands, Formosa, Tibet, Indo-China, and other parts of Central Asia, Equatorial Africa, and Central South America. A special interest attaches to the indigenous products of oceanic islands, i.e., islands separated by a deep sea from any large tract of land. Those who have opportunities could not fail to make interesting discoveries by collecting specimens of the smaller animals (insects, molluscs, etc.,) and plants in these isolated localities. Both in continental countries and on islands the truly indigenous species will have to be sought for on hills and in the remote parts, where they are more likely to have escaped extermination by settlers and the domestic animals introduced by them. In most of the better-known countries the botany has been better investigated than the zoology, and in all these there still remains much to be done in ascertaining the exact station, and the range, both vertical and horizontal, of known species of animals and plants. This leads us to one point which cannot be too strongly insisted on, namely, that some effective means should be adopted by the traveller to record the *exact locality* and *date* of every specimen he collects.*

A traveller may be puzzled, in the midst of the profusion of animal and vegetable forms which he sees around him, to know what to collect and what to leave. Books can be of little service to him on a journey, and he had better at once abandon all idea of encumbering himself with them. A few days study at the principal museums before he starts on his voyage may teach him a great deal, and the cultivation of a habit of close observation and minute comparison of the specimens he obtains will teach him a great deal more. As a general rule all specimens which he may meet with for the first time far in the interior should be preferred to those common near the civilised parts, and he should bear in mind that the few handsome kinds which attract the attention of the natives, and are offered for sale to strangers, are almost sure to be of species well known in European museums. He should strive to obtain as much variety as possible, and not fill his boxes and jars with quantities of specimens of one or a few species, but as some of the rarest and most interesting species closely resemble others which may be more common, he should avail himself of every opportunity of comparing the objects side by side. In most countries, as already remarked, the truly indigenous, and often the rarest, species are to be found only in the mountains at considerable elevations and in the primitive forests, the products of cultivated districts being nearly all widely distributed and well known. In Botany a traveller, if obliged to restrict his collecting, might confine himself to those plants which are remarkable for their economic uses; always taking care to identify the flowers of the tree or shrub whose root, bark, leaves, wood, etc., are used by the natives, and to preserve a few specimens of them.

---

* Cf. British Museum 'Handbook,' p. 47.

But if he has the good fortune to ascend any high mountain not previously explored, he should make as complete a collection of the flowering plants as possible, at the higher elevations. The same may be said of insects found on mountains, where they occur in great diversity—on the shady and cold sides rather than on the sunny slopes—under stones, and about the roots of herbage, especially near springs, on shrubs and low trees, and so forth; for upon a knowledge of the plants and insects of mountain ranges depend the solution of many curious questions regarding the geographical distribution of forms over the earth. In Reptiles, the smaller Batrachians (Frogs, Salamanders, etc.) should not be neglected, especially the extremely numerous family of tree-frogs; the arboreal, or rock-haunting species of Lizards seen out of reach, and the swift-running forms that inhabit sandy plains may be secured with a charge of dust-shot, the saloon pistol being especially handy for this purpose. Snakes should be taken without injuring the head, which is the most important part of the body: a cleft stick may be used in securing them by the neck, or they may be shot, and on reaching camp placed in the jars of spirits. As large a collection as possible should be made of the smaller Fishes and Tortoises of lakes and rivers.

*Mammals and Birds.**—An ordinary geographical expedition will hardly have the means at its disposal for bringing home many specimens of the larger animals, but many species in regions visited only by adventurous explorers are still desiderata in the large museums of Europe; and additional specimens of all genera of which there are numerous closely allied species (e.g., Rodents, Antelope, Deer, etc.), and of all the small nocturnal mammals would be welcome to zoologists. If only portions can be obtained, skulls with horns attached are to be preferred. The smaller birds shot on an excursion should be carried to camp in the game bag, folded in paper, the mouth, anus and any wounds being first plugged with cotton-wool. In a hot climate when the birds have to be carried for some distance before they are skinned, a plug of cotton-wool dipped in a weak solution of formaline or in spirits should be pushed down the gullet into the stomach, before the mouth is plugged with dry cotton-wool. This precaution will insure specimens remaining fresh for many hours.

---

* Cf. British Museum 'Handbook,' pp. 15–32.

Small dull-coloured birds are most likely to be new or interesting.

Immediately after killing a small mammal or bird, make a note of the colour of its eyes and soft parts, and, if time admits, of the dimensions of its trunk and limbs. Full directions for skinning will be found in the British Museum 'Handbook,' pp. 15-29. It should however be mentioned that in large-headed Parrots, Woodpeckers, Ducks, and some other birds, in which the skin of the head cannot be turned back, an incision has to be made in the nape, through which the skull can be pushed and cleansed, the incision being afterwards closed by two or three stitches. In finishing the skin of a bird the feet should be placed side by side, with the claws folded and fastened together by means of a pin run transversely through the soles. The protruding ends of the pin can afterwards be cut off close to the feet. This is Mr. W. Foster's plan, and is by far the best and neatest method. When the skin is dry, it should be laid on its back in one of the trays fitted into the zinc-lined collecting boxes, and secured by means of a couple of stout pins passed through the head at the base of the lower mandible and through the root of the tail. By dovetailing the specimens into one another, they can thus be packed with the least loss of space, and need not again be moved. They require no wrapping or paper, and are much more easily looked over to see that no insects have attacked them.

Preserving Mammals, etc., in Alcohol.—When Mammals cannot be prepared as skins with skulls, according to the British Museum directions, they may prove of service if preserved in spirits. Indeed, when a series of skins has been made, additional specimens may well be placed in spirits, while in the case of Bats half the individuals taken of any species should be skinned and half put in spirits.

On the subject of the preservation of such spirit specimens, the late Dr. G. E. Dobson has published the following 'Hints':—

The general collecting case should be made of strong block tin, or, better still, of copper, rectangular in form, about 2 feet × 1 foot × 1 foot 8 inches in height, having in the top a circular aperture from 6 to 8 inches in diameter, closed by a well-fitting brass screw-cap, the flange of which is made air-tight by a well-greased leather collar. This should fit accurately into a similarly shaped box of inch boards, having a simple flat lid (not projecting beyond the sides), secured by eight long screws, and provided with a strong iron handle.

This case should be filled with the strongest methylated spirits procurable (in foreign countries over-proof rum, brandy, or arrack will suit equally well). If circumstances admit, two or more such cases should be taken, or four wide-mouthed earthenware jars placed in a square wooden case, separated by light wooden partitions, and having their mouths closed by well-fitting bungs tied down with bladder and string. On arrival at the collecting station one of the jars should be half filled with spirits from the tin case. Into this each specimen, as it is obtained, having been slit along the side of the abdomen, should be

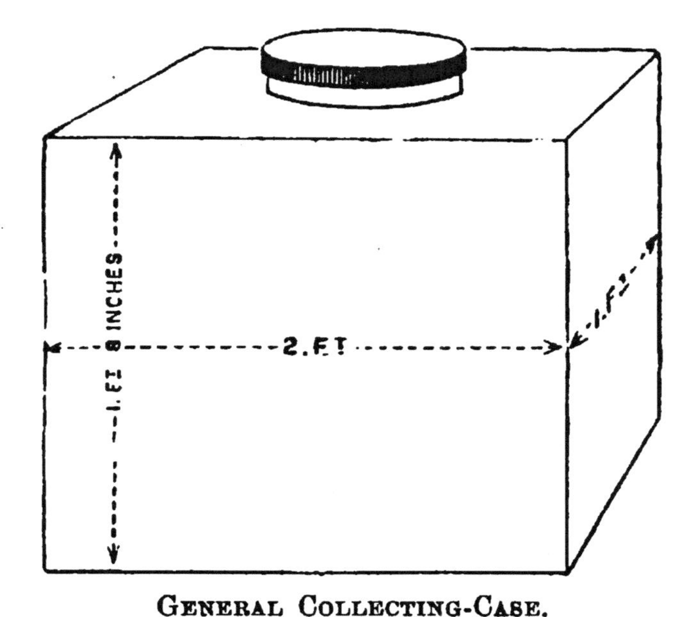

GENERAL COLLECTING-CASE.

put, and allowed to remain 24 hours before being transferred to the general collecting case. When the latter can hold no more the specimens should be removed one by one and packed in the moist state in the other wide-mouthed jars, one above the other, like herrings in a cask, each rolled in a piece of thin cotton cloth, in which a label, having the locality and date written in pencil, should be placed. When the jar has been thus filled to the mouth a glass or two of the strong spirits (kept in reserve) should be poured in so as to fill up the interstices, but not to appear on the surface, which should be covered with a thick layer of cotton-wool. A few drops of carbolic acid, if the spirit be weak, will greatly aid its preserving powers. The bung should then be replaced, secured round the margin outside with a mixture of tallow and wax, and tied down securely with bladder or skin, and the name of the collector and district written legibly outside. The jar is now ready for transmission to any distance, for specimens thus treated will keep good in the vapour alone of strong spirits for months. Other jars may be filled in like manner, and finally, the general collecting case. Incisions should invariably be made in the *sides* (not in the centre line) of all animals, so as to allow the spirits to enter, and no part of the intestines should be removed. In the case of *Tortoises* the opening may be made in the soft parts round the thighs; if this be not done, the body soon becomes distended with gases. *Frogs* should always be first placed in weak spirits, and after being soaked for one or two days, be removed to strong alcohol. *Crabs* should be rolled up alive in thin cotton-cloths, secured by thread tied round, and may then be readily killed by immersion in alcohol; if this be not done they lose many of their limbs in their dying struggles.

Preparation of Skeletons of Animals.—In many cases it will be found impossible to preserve the whole animal, especially if of large size, but it may advantageously be converted into a skeleton by attention to the following directions of the late Sir W. H. Flower, F.R.S.

If the animal is of small size—say not larger than a Fox--take off the skin except from the feet below the wrist and ankle joints. If it is intended to preserve the skin as a zoological specimen as well as the skeleton, the bones of the feet should all be left in the skin; they can be easily extracted afterwards, and will be preserved much more safely in their natural covering. Remove all the contents of the abdominal and thoracic cavities; also the larynx, gullet, and tongue. In doing this be careful to leave attached to the base of the skull the chain of bones which supports the root of the tongue. These may either be left in connection with the skull, or cleaned separately and tied to the skeleton. Then clear away, with the aid of a knife, as much as possible of the flesh from the head, body, and limbs, without cutting or scraping the bones, or separating them from each other. At any intervals that may be necessary during this process it will be desirable, if practicable, to leave the body in water, so as to wash away as much of the blood as possible from the bones, and a few days' soaking in water frequently changed will be an advantage.

The body, with all the bones held in connection by their ligaments, should then be hung up to dry in a place where there is a free current of air, and out of the way of attacks from animals of prey. Before they get hard the limbs may be folded by the side of the body in the most convenient position, or they may be detached and placed inside the trunk.

When thoroughly dry the skeletons may be packed in boxes with any convenient light packing material between them. Each should be well wrapped in a separate piece of paper or canvas, as sometimes insects will attack the ligamentary structures and allow the bones to come apart.

If it can be avoided, skeletons should never be packed up so long as any moisture remains in them, as otherwise decomposition will go on in the still adhering soft parts, and cause an unpleasant smell.

If the animal is of larger size it will be most convenient to take it partially to pieces before or during the cleaning. The head may be separated from the neck, the vertebral column divided into two or more pieces, and the limbs detached from the trunk; but in no case should the small bones of the feet be separated from one another. The parts should then be treated as above described, and all packed together in a canvas bag.

In the cetacea (porpoises, etc.), look for two small bones suspended in the flesh, just below the vertebral column, at the junction of the lumbar and caudal regions (marked externally by the anal aperture). They are the only rudiments of the pelvis, and should always be preserved with the skeleton.

If there is no opportunity of preserving and transporting entire skeletons, the skulls alone may be kept. They should be treated as above described, picked nearly clean, the brain being scooped out through the *foramen magnum*, soaked for a few days in water, and dried.

Every specimen should be carefully labelled with the popular name of the animal, if known, and at all events, with the sex, the *exact locality* at which it was procured, and the *date*.

For the purpose of making entire skeletons, select, if possible, adult animals; but the skulls of animals of all ages may be advantageously collected.

Collectors of skins should always leave the skull intact. It should be taken out, labelled with a corresponding number to that on the skin, and dried with as much flesh on it as possible.

Reptiles and Fishes.—Full directions for preserving these will be found in the British Museum 'Handbook,' pp. 33–47. The following 'hints' were prepared by the late Mr. Osbert Salvin, F.R.S., who collected these animals most successfully in Guatemala:—

Almost any spirit will answer for this purpose, its fitness consisting in the amount of alcohol contained in it. In all cases it is best to procure the strongest possible, as it is less bulky, and water can always be obtained to reduce the strength to the requisite amount. When the spirit sold retail by the natives is not sufficiently strong, by visiting the distillery the traveller can often obtain the first runnings (the strongest) of the still, which will be stronger than he requires undiluted. The spirit used should be reduced to about 20° over proof, and the traveller should always be provided with an alcoholometer. If this is not at hand, a little practice will enable him to ascertain the strength of the spirit from the rapidity with which the bubbles break when rising to the surface of a small quantity shaken in a bottle. When the spirit has been used this test is of no value. When reptiles or fish are first immersed, it will be found that the spirit becomes rapidly weaker. Large specimens absorb the alcohol very speedily. The rapidity with which this absorption takes place should be carefully watched, and in warm climates the liquid tested at least every twelve hours, and fresh spirit added to restore it to its original strength. In colder climates it is not requisite to watch so closely, but practice will show what attention is necessary. It will be found that absorption of alcohol will be about proportionate to the rate of decomposition. Spirit should not be used too strong, as its effect is to contract the outer surface, and close the pores, thus preventing the alcohol from penetrating through to the inner parts of the specimen. *The principal point, then, is to watch that the strength of the spirit does not get below a certain point while the specimen is absorbing alcohol when first put in.* It will be found that after a few days the spirit retains its strength: when this is the case, the specimen will be perfectly preserved. Spirit should not be thrown away, no matter how often used, so long as the traveller has a reserve of sufficient strength to bring it back to its requisite strength.

In selecting specimens for immersion, regard must be paid to the means at the traveller's disposal. Fishes up to 9 inches long may be placed in spirit, after a slit has been cut in the side of the belly to allow the spirit to enter to the entrails. With larger specimens, it is better to pass a long knife outside the ribs, so as to separate the muscles on each side of the vertebræ. It is also as well to remove as much food from the entrails as possible, taking care to leave all these in. The larger specimens can be skinned, leaving, however, the intestines in, and simply removing the flesh. Very large specimens preserved in this way absorb very little spirit. All half-digested food should be removed from snakes and animals. In spite of these precautions, specimens will often appear to be decomposing; but, by more constant attention to re-strengthening the spirit, they can, in most cases, be preserved.

A case (copper is the best), with a top that can be unscrewed and refixed easily, should always be carried as a receptacle. The opening should be large enough to allow the hand to be inserted; this is to hold freshly-caught specimens. When they have become preserved, they can all be removed and soldered up in tin or zinc boxes. Zinc is best, as it does not corrode so easily. The traveller will find it very convenient to take lessons in soldering, and to be able to make his own boxes. If he takes them ready-made, they had best be arranged so as to fit one into another before they are filled. When moving about, all specimens should be wrapped in calico or linen or other rags to prevent their rubbing one against the other. This should also be done to the specimens in the copper case when a move is necessary, as well as to those finally packed for transmission to Europe. These last should have all the interstices between the specimens filled in with cotton-wool or rags. If a leak should occur in a case, specimens thus packed will still be maintained moist, and will keep some time without much injury. Proof spirit should be used when the specimens are finally packed, but it is not necessary that it should be fresh.

*Land and Fresh-water Mollusca.** Full instructions for collecting these will be found in the 'British Museum Handbook,' pp. 113–115. Lieut.-Col. H. H. Godwin-Austen, F.R.S., has contributed the following notes on collecting these animals: Molluscs are always most abundant on limestone rocks, and should be searched for under the larger stones

* Much useful information may be found in the 'Manual of the Mollusca, by S. P. Woodward, F.G.S., one of Weale's series; an admirable book in a small form.

lying about the ground, and under fallen trees and logs in the woods and forests. They may generally be found adhering to the surface of the stone or wood. Many species are often only 0·05 inch in length, so that very close examination is necessary. In damp spots, generally in ravines with a northerly aspect, the dead leaves when damp with dew in the early morning may be turned over one by one, and the under surface examined for minute species; larger species will be found very frequently on the surface of the ground below the layer of decaying vegetable matter. Many may also be secured by tearing off the bark of decaying trees. In the drier parts of the country some species are only to be found among the roots of shrubs, at considerable depth; by digging these up and shaking the earth on to paper, small shells may be found on close examination. At a dry place like Aden, I should expect to find most of the living land-shells in such a habitat. Look well in limestone caves on the damp surface of the rock; some forms hide themselves under a coating of earthy matter. Search also on damp moss and rocks near waterfalls.

Some species will be found high up on the bushes and trees. This is the habit of certain African forms especially; not so in India. A very good idea of the land-shells of a country may at first be obtained by the examination of the beds of the streams, either along the highest flood-line, or in the fine sand and mud collected in the bed. Land-shells found in such situations are usually old and bleached, but the living specimens will not be far off.

The leaves and stems of water plants should be examined, and Confervæ taken out of the water and well washed in a basin; in these, and in the mud of ponds and still rivers, many minute shells may be found.

The best way of preserving minute shells is to put them into glass tubes plugged with wool; it is better than cork. Capital collecting tubes can be made out of the smaller sorts of bamboo and the large grasses. A certain number of every species (at least a dozen) should be preserved in spirits for the sake of the anatomy. It is best to kill them first in water and then put them into spirits; if this is not done they contract, so that it is impossible to form any idea of the form of the mantle and other parts, and they become so hard they are difficult to cut up.

A good method of keeping the small shells and slugs, especially in spirits, is to put them with labels into small tubes plugged with wool, and then place the tubes in a large jar, capable of holding three or four dozen.

Other small shells, ½ to ¾ of an inch in diameter, may be put into pill-boxes at once, for in a dry climate they very soon dry up. The very large animals may be removed by boiling them in water, but when time does not admit of attending to the cleaning of the shells, species, such as Unios, may be put into empty soup-tins and then filled up with dry sand.

It is very important to make a few notes on the colour of the animal, attaching a number for reference on the box or in the tube, and the operculum, when present, should always be preserved.

With respect to slugs, note the surface of the mantle, and always the form of the extremity of the foot, whether pointed or provided with a mucous pore; and again the lobes of the mantle. Preserve them in spirits as above. Drawings from the living animal are invaluable, and should be made if possible. Very little is known of the Asiatic forms; they are of much interest, and have been very little collected.

Insects.—For the best means of preserving the various orders of insects the traveller should consult the 'British Museum Handbook,' pp. 48–89.

Botanical Collecting.—Full instructions will be found in the 'British Museum Handbook,' pp. 116–125.

The following instructions have been prepared by the late Mr. J. Ball, F.R.S.:—

To obtain good specimens of dried plants in a condition serviceable to scientific men, the following are the chief points to be observed:—

1. *Selection of Specimens.*—The object is to give as much information as possible respecting the plant which it is intended to collect. Small plants not exceeding 16 inches in height should be collected *entire with the roots.* Slender plants of greater dimensions may be folded to the same length, and may often be collected entire. Of larger plants, shrubs and trees, the object is to show as much as possible of the plant within the limit of the size of your drying paper. As an universal rule, both the flower and fruit (seed-vessel) should, if possible, be preserved. Of those plants whereon the male and female flowers grow separately, specimens of both should, if possible, be collected.

2. *Conveyance of Specimens to Camp or Station.*—Tin boxes made for the purpose are generally used in Europe for carrying botanical specimens until they can be placed in the drying press. They answer sufficiently well in cool weather, but in hot countries specimens are often partly

withered before they can be laid out; and a rough portfolio, into which the plants can be put when (or soon after) they are gathered, is much to be preferred.

Such a portfolio is easily prepared with two sheets of millboard connected by an endless tape, so as to be easily slung over the shoulder; between these about thirty or forty sheets (60 to 80 folds) of thin soft (more or less bibulous) paper may be carried and kept in place by a strap or piece of twine. With two such portfolios a traveller can carry as many plants as it is possible to collect with advantage in a day. As soon as possible after being gathered, the specimens should be laid roughly between the sheets of paper: except in the case of delicate flowers, no special care is needed, and no harm comes of two or three being put together.

3. *The Drying Press.*—The great object, both to secure good specimens and to save labour and weight of paper, is to get the plants dried quickly,

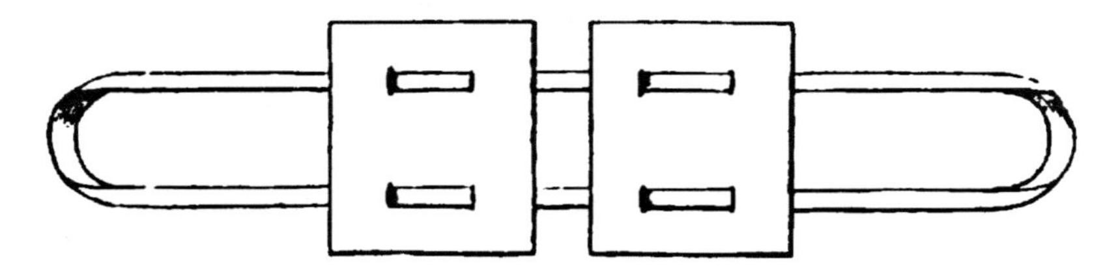

and for this one of the first conditions is to lose as little time as possible. When practicable, the specimens should always be put in the press on the same day on which they are gathered. The press should be made with two outer gratings of iron wire; the outer frame of strong wire, about a quarter of an inch in diameter—the size being that of the paper used. Between these the paper is laid. As to the choice of drying paper, the general rule is, that the coarser it is the better, provided it be quite or nearly quite free from size.

To enable the plants to dry quickly, the traveller should be provided with light wooden gratings of the same size as the drying paper. I think the size 18 inches × 12 inches is quite large enough. The iron wire outer gratings may with advantage be a quarter of an inch longer and broader to save the edges of the wooden gratings.

These should be made of light laths fastened with a few nails (all the better if these are of copper), the interstices should be rather less than.

three-quarters of an inch, at all events not more. Their use is to allow the air to circulate through the pile of plants that are being dried. One should be inserted at each interval of about two inches (counting the drying paper and the plants laid out for drying), and when this is done the parcel may with advantage be exposed to the sun or placed near a fire, as the case may be. In dry warm climates, the majority of plants

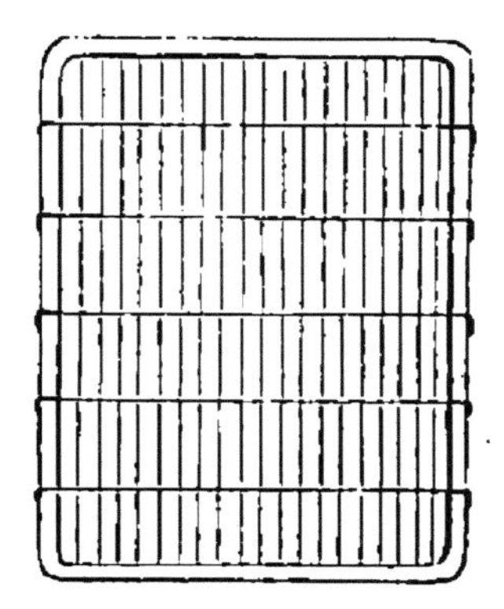

may be dried in the course of a few days, and will be fit to pack up, without any need of changing the drying paper in which they were originally placed; but in damp weather, and in regard to plants of thick fleshy foliage, it is usually necessary to change the paper more than once before the specimens are thoroughly dry.

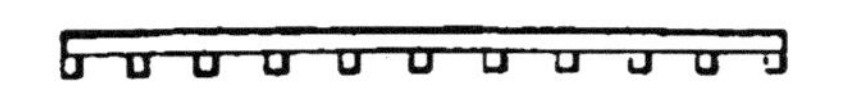

GRATING SEEN FROM THE EDGE.

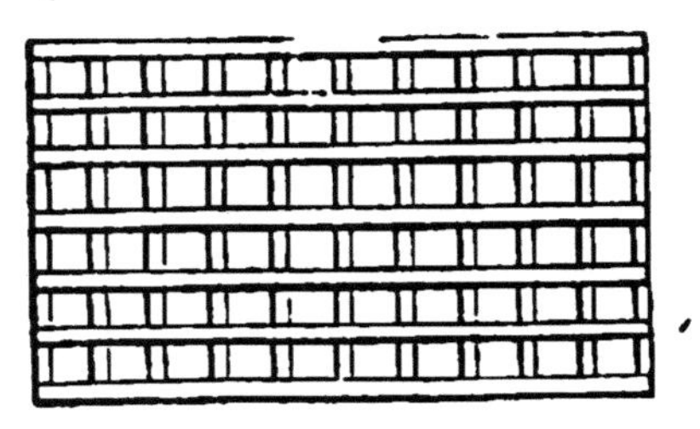

GRATING SEEN FROM ABOVE.

The pile of paper, with plants between each five or six thicknesses of paper, and gratings at intervals of about two inches, should be squeezed between the outer (iron) gratings by means of two strong straps. Too much pressure is not desirable. For a pile ten or twelve inches thick, the parcel may be pulled nearly as tight as a moderate man can do it; but in proportion as the thickness is less, the pressure should be moderated.

Plants with fleshy leaves are very difficult to dry well. The best way is to dip them in quite boiling water for a minute or less, then to lay them between a few sheets of drying paper with slight pressure, merely to remove the exterior moisture, and then place them (when externally dry) in the drying press. Plants collected in rain should be treated in a similar way to remove outer moisture before it is attempted to dry them.

4. When once dry, plants may be packed away between paper of almost any kind. Old newspapers answer very well. The only precaution needed is to preserve them from insects.

The chief trouble in collecting plants is to get the paper already used thoroughly dry before it is again employed. The best resource in dry climates is to stretch cords and hang these papers exposed to sun and air. Artificial heat must be resorted to in wet seasons, but the process is then slow and troublesome.

For a traveller wishing to make large collections, the time consumed in changing the paper in which the plants are dried becomes an important consideration. I have adopted with advantage a suggestion of the late Professor A. Gray to use, instead of ordinary drying paper, sheets cut to the proper size, of the paper-felt which is used for laying under carpets. The specimens when originally laid out for drying are placed within sheets of thin paper without size, such as filtering paper, and as a rule these do not need to be changed. One sheet of felt-paper is generally sufficient between each layer of plants, and the operation of changing the paper is very quickly effected.

It is an important rule to note *the locality* where the specimens have been collected, *with the date.* If proper care be taken to keep together all the specimens collected at the same time, it is not necessary to place a separate scrap of paper within each sheet; but it is advisable to do this when the dried specimens are packed for transmission home.

5. *Seeds.*—Travellers may easily make valuable contributions to our knowledge of the vegetation of distant countries by preserving seeds of remarkable and unusual plants. The only precautions necessary are, to select seeds that are fully ripe; if enclosed in a seed vessel, or covering of a succulent character, to take care that this is thoroughly dried before they are packed; and that they are preserved from moisture during the homeward voyage. Small seeds may be enclosed in paper, the larger kinds in canvas bags, and the whole wrapped in a piece of oiled cloth,

It is very desirable to keep each description of seed separate, and to note the place where it was gathered, with indications of altitude, soil, and climate.

6. *Bulbs.*—These are easily obtained, but as a rule, they should be taken only at the end of the growing season, and kept until the leaves are quite withered. They should be packed dry in a small box with shavings, or other elastic stuffing. The same treatment will suit the pseudo-bulbs of some orchids.

7. *Fleshy Tubers.*—Fleshy and thick tubers are best sent in boxes, wrapped in slightly moist materials, such as cocoa-nut fibre, peat, or leaf mould.

8. *Living Plants.*—As a general rule, these require to be established in pots or boxes for some time before being packed for transmission. They travel best in what are called Wardian cases; but an ordinary wooden box covered with a glass top, and with sufficient moisture in the soil and air to prevent excessive evaporation, is found to answer the purpose. The cases should be kept on deck under some protection from the direct heat of the sun. Tropical plants should be despatched so as to reach England during the summer months. At other seasons they are liable to perish from cold.

9. *Succulent Plants*, such as cacti, aloe, houseleeks, &c., survive for a long time if packed without earth in a perfectly dry box, with sufficient openings for ventilation.

10. Small plants with woody roots and cuttings of larger species of plants from the north or south temperate zones often travel successfully when merely packed with a little soil, slightly moist, about the roots, and a wrapping of damp moss, or similar substance, tied up in thick paper or canvas. There is, however, much risk of failure in these cases where, on the homeward voyage, it is necessary to pass through the tropics.

As a general rule, plants are more often injured by excess of moisture than by being sent too dry.

It is desirable to make use of every favourable opportunity for sending botanical collections of all kinds to England, as in hot countries they are always exposed to risk of injury.

It is scarcely necessary to mention that living plants, as well as seeds, and bulbs, should be placed in the hands of skilful gardeners after reaching this country. The chance of preserving interesting specimens is commonly much greater when they are sent to botanic gardens than when entrusted to private cultivators. In all cases information as to the soil and climate of the native home of the plant is a necessary guide to proper treatment.

Fossils.—The collection of fossils and minerals (except in the case of the discovery of new localities for valuable metals) is not to be recommended to the traveller, if he is not a geologist. Fossils from an unexplored country are of little use unless the nature and order of superposition of the strata in which they are found can be at the same time investigated. In the cases, however, of recent alluvial strata or the supposed beds of ancient lakes, or deposits in caves, or raised sea-beaches containing shells or bones of vertebrate animals, the traveller will do well to bring away specimens if a good opportunity offers. If the plan of the expedition includes the collection of fossil remains, the traveller will, of course, provide himself with a proper geological outfit, and obtain the necessary instructions before leaving Europe. (*See* Section III.) For suggestions as to the collecting and preserving fossils and minerals, cf. British Museum 'Handbook,' pp. 126–135.

General Remarks.—All collections made in tropical countries should be sent to Europe with the least possible delay, as they soon become deteriorated and spoilt unless great care be bestowed upon them.

Observations of Habits, etc.—Travellers have excellent opportunities of observing the habits of animals in a state of nature, and these 'Hints' would be very deficient were not a few remarks made upon this subject. To know what to observe in the economy of animals is in itself an accomplishment which it would be unreasonable to expect the general traveller to possess, and without this he may bring home only insignificant details, contributing but little to our stock of real knowledge. One general rule, however, may be kept always present to the mind, and this is, that anything concerning animals which bears upon the relations of species to their conditions of life is well worth observing and recording. Thus, it is important to note the various enemies which each species has to contend with, not only at one epoch in its life, but at every stage from birth to death, and at different seasons and in different localities. The way in which the existence of enemies limits the range of a species should also be noticed. The inorganic influences which inimically affect species, especially intermittently (such as the occurrence of disastrous seasons), and which are likely to operate in limiting their ranges, are also important subjects of inquiry. The migrations of animals, and especially any facts about the irruption of species into districts previously uninhabited by them, are well worth recording. The food of each species should be noticed, and if any change of customary food is observed, owing to the failure of the supply, it should be carefully recorded. The use in nature of any peculiar physical conformation of animals, the object of ornamentation, and so forth, should also be investigated whenever opportunity occurs. Any facts relating to the interbreeding in a state of nature of allied varieties, or the converse—that is, the antipathy to intercrossing of allied varieties—would be extremely interesting. In short, the traveller should bear in mind that facts having a philosophical bearing are much more important than mere anecdotes about animals.

To observe the actions of the larger animals, a telescope or field-glass will be necessary. The traveller should bear in mind, if a microscope is needed in his journey, that by unscrewing the tubes of the telescope in which all the small glasses are contained, a compound microscope of considerable power may be produced.

V.

ANTHROPOLOGY.

By E. B. TYLOR, D.C.L., F.R.S.

THE characters of men's bodies and minds being matters of common observation, Europeans not specially trained in anthropology, who have happened to be thrown among little-known tribes, often bring home valuable anthropological information. Though explorers, traders, and colonists have made their way into almost every corner of the earth, it is surprising to find how many new facts may still be noted down by any careful observer. If familiar with anthropological methods, he will, of course, observe more and better. The hints here given will serve to draw attention to interesting points which might otherwise be overlooked. Directions for such investigation, drawn up in much greater detail, will be found in the small British Association manual entitled: 'Notes and Queries on Anthropology' (Anthropological Institute, 3, Hanover Square, W.).

Physical Characters.—On first coming among an unfamiliar race, such as the Negroes, the traveller is apt to think them almost alike, till after a few days he learns to distinguish individuals more sharply. This first impression, however, has a value of its own, for what he vaguely perceived was the general type of the race, which he may afterwards gain a more perfect idea of by careful comparison. Among tribes who for many generations have led a simple uniform life and mixed little with strangers, the general likeness of build and feature is very close, as may be seen in a photograph of a party of Caribs or Andamaners, whose uniformity contrasts instructively with the individualised faces of a party of Europeans. The consequence is that a traveller among a rude people, if he has something of the artist's faculty of judging form, may select groups for photography which will fairly represent the type of a whole tribe or nation. While such portrait-groups are admirable for giving the general idea of a race, characteristic features belonging to it should be treated separately. For instance, to do justice to the Tartar eye or the Australian forehead, the individual feature must be carefully sketched or photographed large.

How deceptive mere unmeasured impressions of size may be is shown by the well-known example of the Patagonians, who, though really only tall men (averaging 5 feet 11 inches), long had the reputation of a race of giants. Such measurements as any traveller can take with a measuring-tape and a three-foot rule with sliding square are good if taken with proper precautions. As the object of the anthropologist is to get a general idea of a race, it may be in some respects misleading to measure at random one or two individuals who are perhaps not fair specimens. If only a few can be measured, they should be selected of ordinary average build, full-grown but not aged. What is much better is to measure a large number (twenty to fifty) of persons taken indiscriminately as they come, and to record the measurements of each with sex, age, name, locality, etc. Such a table can afterwards be so classified as to show not only the average or mean size, but the proportion of persons who vary more or less from that mean size; in fact, it represents on a small scale the distribution of stature, etc., in the whole people. Gigantic or dwarfish individuals, if not deformed, are interesting, as showing to what extremes the race may run. The most ordinary measurements are height, girth round chest, fathom or length of outstretched arms, length of arm from shoulder and leg from hip, length of hand and foot. The traveller may find that such measuring of another race shows very different stature and girth from that of his own companions, who, if they are well-grown Europeans, may stand 5 feet 8 inches to 6 feet, and measure 34 to 36 inches round the chest. Beyond this, he will find that the relative proportions of parts of the body differ from those he is accustomed to. An example of this is seen by placing Europeans and negroes side by side, and noticing how much nearer the knee the negro's finger-tips will reach. It will be found that body measurement needs skill in taking the corresponding points, and in fact all but the simplest measures require some knowledge of anatomy. This is especially the case with skull measurements. There are instruments for taking the dimensions of the living head, and with care and practice the untrained observer may get at some of the more conspicuous, such as the relative length and width of the skull as taken by hatters. This roughly indicates the marked difference between dolichocephalic or long-headed peoples, like the African negro, and brachycephalic or short-headed peoples, like the Kalmuks and other Tartars. Attention should be paid also to the degree of prognathism or projection of jaw, which in some races, as the Australian, gives a "muzzle" unlike the English type. Where practicable, native skeletons, and especially skulls, should be sent home for accurate examination. How far this can be done depends much on the feeling of the people; for while some tribes do not object to the removal of bones, especially if not of their own kinsfolk, in other districts it is hardly safe to risk the displeasure of the natives at the removal of the dead—a feeling which is not only due to affection or respect, but even more to terror of the vengeance of the ghosts whose relics have been disturbed.

In describing complexion, such terms as "brown" or "olive," so often used without further definition in books of travel, are too inexact to be of use. Broca's scale of colours (*see* the Anthropological 'Notes and Queries') gives means of matching the tints of skin, hair and eyes; if this is not forthcoming, the paint-box should be used to record them. Among rude tribes, the colour of the skin is often so masked by paint and dirt that the subject must be washed to see the real complexion. Hair is also an important race-mark, varying as it does in colour from flaxen to black, and also in form and size of the hairs; for instance the American Indian's coarse straight hair seems almost like a horse's tail in comparison with the Bushman's hair with its natural frizz of tiny spirals. Locks of hair should therefore be collected. The traveller, however, will often find some difficulty in getting such specimens, from the objection prevalent in the uncivilised world of letting any part of the body, such as hair and nail-clippings, pass into strangers' hands lest they should be used to bewitch their former owner. Even in such countries as Italy, to ask for a lock of a peasant-girl's hair may lead to the anthropologist being suspected of wishing to practise love-charms on her.

Differences of temperament between nations are commonly to be noticed; for instance, in comparing the shy and grave Malays with the boisterous Africans. It is an interesting but difficult problem how far such differences are due to inherited race-character, and how far to such social influences as education and custom, and to the conditions of life being cheerful or depressing. Nor has it yet been determined how far emotions are differently expressed by different races, so that it is worth while to notice particularly if their smiling, laughing, frowning, weeping, blushing, etc., differ perceptibly from ours. The acuteness of the senses of sight, hearing and smell, among wild peoples is often remarkable, but this subject is one on which many accounts have been given which require sifting. The skill of savages in path-finding and tracking depends in great measure on this being one of their most necessary arts of life to which they are trained from childhood, as, in an inferior degree, gipsies are with us. The native hunter or guide's methods of following the track of an animal, or finding his own way home by slight signs, such as bent twigs, and keeping general direction through the forest by the sky and the sheltered sides of the trees, are very interesting, though when learnt they lose much of their marvellous appearance. The testing of the mental powers of various races is an interesting research, for which good opportunities now and then occur. It is established that some races are inferior to others in volume and complexity of brain, Australians and Africans being in this respect below Europeans, and the question is to determine what differences of mind may correspond. Setting aside the contemptuous notions of uneducated Europeans as to the minds of "black-fellows" or "niggers," what is required is, to compare the capacity of two races under similar circumstances. This is made difficult by the fact of different training. For instance, it would not be fair to compare the European sportsman's skill in woodcraft and hunting with that of the native hunter, who has done nothing else since childhood; while, on the other hand, the European, who has always lived among civilised people, owes to his education so much of his superior reasoning powers, that it is mostly impossible to get his mind into comparison with a savage's. One of the best tests is the progress made by native and European children in colonial or missionary schools, as to which it is commonly stated that children of African or American tribes learn as fast as or faster than European children up to about twelve, but then fall behind. Even here it is evident that other causes besides mental power may be at work, among them the discouragement of the native children when they become aware of their social inferiority. The subject is one of great importance, both scientifically and as bearing on practical government.

Both as a matter of anthropology and of practical politics the suitability of particular races to particular climates is of great interest; sometimes this depends on one race being free from a disease from which another suffers, as in the well-known immunity of negroes from yellow fever. Or it may be evident that tribes have become acclimatised, so as to resist influences which are deadly to strangers; for instance, the Khonds flourish in the hills of Orissa, where not only Europeans but the Hindus of the plains sicken of the malaria in the unhealthy season. That such peculiarities of constitution are inherited and pass into the nature of the race, is one of the keys to the obscure problem of the origins of the various races of man as connected with their spread over the globe. As yet this problem has not passed much beyond the stage of collecting information, and no pains should be spared to get at facts thus bearing on the history and development of the human species. European medical men in districts inhabited by uncivilised races have often made important observations of this kind, which they are glad to communicate, though being occupied with professional work they do not follow them up. In all races there occur abnormal varieties, which should be observed with reference to their being hereditary, such as Albinos, whose dead-whiteness is due to absence of pigment from the skin. Even such tendencies as that to the occurrence of red hair where the ordinary hue is black, or to melanism or diseased darkening of the skin, are worth remark. It is essential to discover how far these descend from parents to children, which is not the case with such alterations as that of the Chinese feet, which, in spite of generations of cramping, continue of the natural shape in the children.

Language.—Before coming to actual language, remark may be made on the natural communication of all races carried on by pantomimic signs without spoken words. This is the "gesture language" to which we are accustomed among the deaf-and-dumb, and which sometimes also comes into practical use between tribes ignorant of one another's languages, as on the American prairies. It is so far the same in principle everywhere, that the explorer visiting a new tribe, having to make frequent use of signs to supplement his interpreter, or to eke out his own scanty knowledge of the native language, soon adapts himself to the particular signs in vogue. He will observe that, as to most common signs, such as asking for food or drink, or beckoning or warning off a stranger, he understands and is understood quite naturally. Signs which are puzzling at first sight will prove on examination to be intelligible. Some are imitative gestures cut short to save trouble, or they may have a meaning which was once evident, like the American Indian sign for dog, made by trailing two forked fingers, which does not show its meaning now, but did so in past times, when one of the principal occupations of the dog was to trail a pair of tent-poles attached to his back. Besides its practical use, the gesture-language has much scientific interest from the perfect way in which it exposes the working of the human mind, expressing itself by a series of steps which are all intelligible. It will be particularly observed that it has a strict syntax; for instance, that the quality or adjective must always follow the subject or substantive it is applied to. Thus, "the white box" may be expressed by imitating the shape and opening of a box, and then touching a piece of linen or paper to show its colour; but if the signs be put in the contrary order, as in the English words, the native will be perplexed. It is worth while, in countries where gesture-language is regularly used, to note down the usual signs and their exact order.

In recording a vocabulary of a language not yet reduced to form in a grammar and dictionary, the traveller may seek for equivalents of the principal classes of words in his own grammar: verbs, substantives, adjectives, pronouns, prepositions, etc. But the structure of the language he is examining will probably differ from any he is familiar with, the words actually used not coming precisely into these classes. The best method is for the traveller to learn a simple sentence, such as "the men are coming," and to ascertain what changes will convert them into "the men are going," "the women are coming." He thus arrives at the real elements of the language and the method of combining them. Having arrived at this point, he will be able to collect and classify current ideas, such as the following:—

Actions—as stand, walk, sleep, eat, see, make, etc.

Natural Objects and Elements—as sun, moon, star, mountain, river, fire, water, etc.

Man and other Animals—as man, woman, boy, girl, deer, buck, doe, eagle, eagles, etc.

Parts of Body—as head, arm, leg, skin, bone, blood, etc.

Trees and Plants.

Numerals (noticing how far they extend, and whether referring to fingers).

Instruments and Appliances—as spear, bow, hatchet, needle, pot, boat, cord, house, roof, &c.

Arts and Pastimes—as picture, paint, carving, statue, song, dance, toy, game, riddle, &c.

Family Relationships (as defined by native custom).

Social and Legal Terms—as chief, freeman, slave, witness, punishment, fine, &c.

Religious Terms—as soul, spirit, dream, vision, sacrifice, penance, &c.

Moral Terms—as truth, falsehood, kindness, treachery, love, &c.

Abstract Terms, relating to time, space, colour, shape, power, cause, &c.

The interjections used in any language can be noted, whether they are organic expressions of emotion, like *oh! ugh! ur-r-r!* or sounds the nature of which is not so evident. Also imitative words which name animals from their cries, or express sounding objects or actions by their sounds, are common in all languages, and strike the stranger. Examples of such are *kah-kah* for a crow, *twonk* for a frog, *pututu* for a shell-trumpet, *haitschu* for to sneeze. When such imitative words are noticed passing into other meanings where the connection with sound is not obvious, they become interesting facts in the development of language; as, to take a familiar example from English, the imitative verb to *puff* becomes a term for light pastry and metaphorically blown-up praise.

It is only when the traveller has a long or close acquaintance with a tribe, that he is able to deal satisfactorily with the vocabulary and structure of their language. To be able to carry on a conversation in broken sentences is not enough, for an actual grammar and dictionary is required to enable philologists to make out the structure and affinities with other languages. It used to be customary to send out English lists of thirty or forty ordinary words to have equivalents put to them in native languages. As every detail of this kind is worth having, these lists cannot be said to be quite worthless, but they go hardly any way toward what is really wanted. They are liable to frequent mistakes, as when the barbarian, from whom the white man is trying to get the term "foot," answers with a word meaning "my leg," which is carefully taken down and printed. Such poor vocabularies cannot even be relied on to show whether a language belongs to a particular family, for the very word which seems to prove this may be borrowed. Thus, in various African vocabularies, there appears the word *sapun* (or something similar) with the meaning of *soap;* but this is a Latin word which has spread far and wide from one country to another, and proves nothing as to original connection between languages which have adopted it. While it is best not to under-rate the difficulty of collecting such information as to a little-known dialect as will be really of service to philology, it must be remembered that travellers still often have opportunities of preserving relics of languages, or at any rate special dialects, which are on the point of dying out unrecorded. Where no proper grammar and dictionary has been compiled, it is often possible to find some European or some interpreter fairly conversant with the language, with whose aid a vocabulary may be written out and sentences analysed grammatically, which, when read over to intelligent natives and criticised by them, may be worked into good linguistic material. It is worth while to pay attention to native names of plants, minerals, &c., as well as of places and persons, for these are often terms carrying significant meaning. Thus *ipecacuanha* is stated by Martius to be *i-pe-caa-guéne*, which in the Tupi language of Brazil, signifies "the little wayside plant which makes vomit."

Arts and Sciences.—The less civilised a nation is, the ruder are its tools and contrivances; but these are often worked with curious skill in getting excellent results with the roughest means. Stone implements have now been so supplanted by iron that they are not easily found in actual use. If a chance of seeing them occurs, as, for instance, among some Californian tribe, who still chip out arrow-heads of obsidian, it is well to get a lesson in the curious and difficult art of stone-implement making. In general, tools and implements differing from those of the civilised world, even down to the pointed stick for root-digging and planting, are worth collecting, and to learn their use from a skilled hand often brings into view remarkable peculiarities. This is the case with many cudgel or boomerang-like weapons thrown at game, slings or spear-throwers for hurling darts to greater distances than they can be sent by hand, blow-tubes for killing birds, and even the bow-and-arrow, which in northern Asia and America shows the ancient Scythian or Tartar form, having to be bent inside out to string it. Though fire is now practically made almost everywhere with flint and steel or lucifers, in some districts, as South Africa or Polynesia, people still know the primitive method of fire-making by rubbing or drilling a pointed stick into another piece of wood. Europeans find difficulty in learning this old art, which requires some knack. As is well known to sportsmen, different districts have their special devices for netting, trapping and other ways of taking game and fish, some of which are well worth notice, such as spearing or shooting fish under water, artificial decoys, and the spring-traps set with bent boughs, which are supposed to have first suggested the idea of the bow. While the use of dogs in hunting is found in most parts of the world, there is the utmost variety of breeds and training. Agriculture in its lower stages is carried on by simple processes; but interesting questions arise as to the origin of its grain and fruits, and the alterations in them by transplanting into a new climate and by ages of cultivation. Thus in Chili there is found wild what botanists consider the original potato; but while maize was a staple of both Americas at the time of Columbus, its original form has no more been identified than that of wheat in the Old World. The cookery of all nations is in principle known to the civilised European; but there are special preparations to notice, such as bucaning or drying meat on a hurdle above a slow fire, broiling kebabs or morsels of meat on the skewer in the East, etc. Many peoples have something peculiar in the way of beverages, such as the chewed Polynesian *kava*, or the South American *maté* sucked through a tube. Especially fermented liquors have great variety, such as the *kumiss* from mares' milk in Tartary, the *pombe* or millet-beer of Africa, and the *kvass* or rye-beer of Russia. The rudest pottery made by hand, not thrown on the wheel, is less and less often met with, but ornamentation traceable to its being moulded on baskets is to be seen; and calabashes, joints of bamboo, and close-plaited baskets are used for water-vessels, and even to boil in. Among the curious processes of metal-working, contrasting with those of modern Europe, though often showing skill of their own, may be mentioned the simple African smelting-forge by which iron-ore is reduced with charcoal in a hole in the ground, the draught being supplied by a pair of skins for bellows. In the far East a kind of air-pump is used, of which the barrels are hollowed logs. The Chinese art of patching cast-iron with melted metal surprises a European, and the Hindu manufacture of native steel (*wootz*) is a remarkable process. No nation now exists absolutely in the Bronze Age, but this alloy still occupies something of its old place in Oriental industry. As an example of the methods still to be seen, may be mentioned the Burmese bell-founding, which is done, not in a hollow mould of sand, but by what in Europe is called the *cire perdue* process, the model of the bell being made in bees-wax and imbedded in the sand-mould, the wax being melted and the hot metal taking its place. The whole history of machinery is open to the traveller, who still meets with every stage of its development, from savagery upward. He sees, for instance, every tilling implement from the stake with fire-hardened point, and the hoe of crooked branch, up to the modern forms of plough. In like manner he can trace the line from the rudest stone-crushers or rubbers for grinding seed or grain up to the rotating hand-mills or querns still common in the East, and surviving even in Scotland. From time to time some special contrivance may be seen near its original home, as in South America the curious plaited tube for wringing out the juice from cassava, or the net hammock which still retains its native Haitian name *hamaca*. Architecture still preserves in different regions interesting early stages of development, from the rudest breakwinds, or beehive huts of wattled boughs, up to houses of logs and hewn timber, structures of mud and adobes, and masonry of rough or hewn stone. Even the construction of the bough-hut or the log-house often has its peculiarities in the arrangements of posts and rafters. Among the modes of construction which interest the student of architectural history is building with rough unhewn stones. Many examples of "rude stone monuments" are to be seen on our own moors and hills. The most familiar kinds are *dolmens* (*i.e.* "table-stones"), formed by upright stones bearing a cap-stone; they were burial-places, and analogous to the cists or chambers of rough slabs within burial-mounds. Less clearly explicable are the single standing-stones or *menhirs* (*i.e.* "long-stones"), and the circles of stones or *cromlechs*. Ancient and obscure in meaning as such monuments are in Europe, there are regions where their construction or use comes down to modern times, especially in India, where among certain tribes the deposit of ashes of the dead in dolmens, the erection of menhirs in memory of great men, and even sacrifice in stone circles, are well-known customs. The traveller may also sometimes have opportunities of observing the ancient architectural construction by fitting together many-sided stones into what are sometimes called Cyclopean walls, a kind of building which seems to have preceded the use of squared blocks, fastened together with clamps or with mortar. Vaulting or roofing by means of courses of stones projecting inwards one course above the other (much as children build with their wooden bricks), so as to form what architects call a "false arch," is an ancient mode of construction found in various parts of the world where the "true arch" with its keystone has not superseded it. It often appears that rude nations have copied the more artistic buildings of higher neighbours, or inherited ancient architectural traditions. Thus traces of Indian architecture have found their way into the islands of the Eastern Archipelago, and hollow squares of mud-built houses round a courtyard in northern Africa have their plan from the Asiatic caravanserai. In boat-building some primitive forms, as the "dug-out," hollowed by the aid of fire from a tree-trunk, and the bark-canoe, are found in such distant regions that we cannot guess where they had their origin. When, however, it comes to the outrigger-canoe, this belongs to a district which, though very large, is still limited, so that we may at least guess whereabouts it first came into use, and it is important to note every island to which it has since travelled. So there is much in the peculiar build and rig of Malay prahus, Chinese junks, etc., which is worth noting as part of the history of ship-building. This may suffice to give a general idea of the kind of information as to the local arts which it is worth while to collect, and to illustrate by drawings and photographs of objects too large to bring away.

Naturally, nations below the upper levels of culture have little or no science to teach us, but many of their ideas are interesting as marking stages in the history of the human mind. Thus, in the art of counting, which is one of the foundations of science, it is common to find the primitive method of counting by fingers and toes still in practical use, while in many languages the numeral words have evidently grown up out of such a state of things. Thus *lima*, the well-known Polynesian word for five, meant "hand," before it passed into a numeral. All devices for counting are worth notice, from the African little sticks for units and larger sticks for tens, up to the ball-frames with which the Chinese and Russian traders reckon so rapidly and correctly. It is a sign of lowness in a tribe not to use measures and weights, and where these appear in a rough way, it is interesting to discover whether vague lengths, such as finger, foot, pace, are used, or whether standard measures and weights have come in. If so, these should be estimated according to our standards with as much accuracy as possible, as it may thus become possible to ascertain their history. In connection with this comes the question of money, as to whether commerce is still in the rudimentary stage of exchanging gifts, or has passed into regular barter, or risen to regular trade, with some sort of money to represent value, even if the circulating medium be only cowries, or bits of iron, or cakes of salt, all which are current money to this day in parts of Africa. Outside the present higher civilisation, more or less primitive ideas of astronomy and geography will be found to prevail. Among tribes like the American Indians the obvious view suggested by the senses still prevails, that the earth is a flat round disc (or sometimes square, with four quarters or winds) overarched by a solid dome or firmament, on which the sun and moon travel—in inland countries going in and out at holes or doors on the horizon, or, if the sea bounds the view, rising from and plunging into its waves at sunrise and sunset. These early notions are to us very instructive, as they enable us to realise the conceptions of the universe which have come down to us in the ancient books of the world, but which scientific education has uprooted from our own minds. With these cosmic ideas are found among the lowest races the two natural periods of time, namely, the lunar month and the solar year, determined by recurring winters, summers, or rainy seasons. Such tribes divide the day roughly by the sun's height in the sky, but among peoples civilised enough to have time-measures and the sun-dial, there is a tolerably accurate knowledge of the sun's place at the longest and shortest days, and indeed, throughout the year. The astronomy of such countries as India has been of course described by professional astronomers; but among ruder nations there is still a great deal unrecorded—for instance, as to the constellations into which they map out the heavens. This likening stars and star-groups to animals and other objects is almost universal among mankind. Savages like the Australians still make fanciful stories about them, as that Castor and Pollux are two native hunters, who pursue the kangaroo (Capella) and kill him at the beginning of the hot season. Such stories enable us to understand the myths of the Classical Dictionary, while modern astronomers keep up the old constellations as a convenient mode of mapping out the sky. As to maps of the earth, even low tribes have some notion of their principle, and can roughly draw the chart of their own district, which they should be encouraged to do. Native knowledge of natural history differs from much of their rude science in its quality, often being of great positive value. The savage or barbarian hunter knows the animals of his own region and their habits with remarkable accuracy, and inherited experience has taught him that certain plants have industrial and medicinal uses. Thus, in South America the Europeans learnt the use of India-rubber or caoutchouc, which the native tribes were accustomed to make into vessels and playing-balls, and of the Peruvian bark or cinchona, which was already given to patients in fever.

Here a few words may be said of magic, which, though so utterly futile in practice, is a sort of early and unsuccessful attempt at science. It is easy, on looking into the proceedings of the magician, to see that many of them are merely attempts to work by false analogy or deceptive association of ideas. The attempt to hurt or kill a person by cutting or piercing a rude picture or image representing him, which is met with in all the four quarters of the globe, is a perfect example of the way in which sorcerers mistake mere association of ideas for real cause and effect. Examined from this point of view, it will be found that a large proportion of the magic rites of the world will explain their own meaning. It is true that this is not the only principle at work in the magician's mind; for instance, he seems to reason in a loose way that any extraordinary thing will produce any extraordinary effect, so that the peculiar stones and bits of wood which we should call curiosities become to the African sorcerer powerful fetishes. It will often be noticed that arts belonging to the systematic magic of the civilised world, which has its source in Babylon and Egypt, have found their way into distant lands more readily indeed than useful knowledge, so that they may even be met with among barbaric tribes. Thus it has lately been pointed out that the system of lucky and unlucky days, which led the natives in Madagascar to kill many infants as of inauspicious birth, is adopted from Arabic magic, and it is to be expected that many other magical arts, if their formulas are accurately described, may in like manner be traced to their origin.

Society.—One of the most interesting features of savage and barbaric life is the existence of an unwritten code of moral conduct, by which families and tribes are practically held together. There may be no laws to punish crime, and the local religion may no more concern itse directly with men's behaviour to one another than it did in the South Sea Islands. But among the roughest people there is family affection, and some degree of mutual help and trust, without which, indeed, it is obvious that society would break up, perhaps in general slaughter. Considering the importance of this primitive morality in the history of mankind, it is unfortunate that the attention of travellers has been so little drawn to it, that our information is most meagre as to how far family affection among rude tribes may be taken to be instinctive, like that of the lower animals, or how far morality is produced by public opinion favouring such conduct as is for the public good, but blaming acts which do harm to the tribe. It is desirable to inquire what conduct is sanctioned by custom among any people, whether, for instance, infanticide is thought right or wrong, what freedom of behaviour is approved in youths and girls, and so on. For though breaches of custom may not be actually punishable, experience will soon convince any explorer among any rude tribe that custom acts in regulating their life even more strictly than among ourselves. The notion of even savages leading a free and unrestrained life is contradicted by those who know them best; in fact, they are bound in every act by ancestral custom. While each tribe thus has its moral standard of right and wrong, this differs much in different tribes, and one must become intimately acquainted with any people to ascertain what are really their ruling principles of life. Accounts have been often given of the natural virtue and happiness of rude tribes, as in the forests of Guiana or the hills of Bengal, where the simple native life is marked by truthfulness, honesty, cheerfulness, and kindness, which contrast in a striking way with the habits of low-class Europeans. There are few phenomena in the world more instructive than morality thus existing in practical independence either of law or religion. It may still be possible to observe it for a few years before it is altered by contact with civilisation, which, whether it raises or lowers on the whole the native level, must supersede in great measure this simple family morality.

The unit of social life is the family, and the family is based on a marriage-law. Travellers who have not looked carefully into the social rules of tribes they were describing, or whose experience has been of tribes in a state of decay, have sometimes reported that marriage hardly existed. But this state of things is not confirmed as descriptive of any healthy human society, however rude; in fact, the absence of definite marriage appears incompatible with the continued existence of a tribe. Therefore statements of this kind made by former visitors should be carefully sifted, and marriage-laws in general deserve careful study. The explorer will hardly meet with marriage at so low a stage that the union can be described as little beyond annual pairing; but where divorce is almost unrestricted, as in some African tribes, there is more or less approach to this condition, which is possible, though unusual, under such laws as that of Islam. Polygamy, which exists over a large part of the globe, is a well-understood system, but information is less complete as to the reasons which have here and there led to its opposite polyandry, as among the Toda hill-tribes and the Nairs in South India. Among customs deserving inquiry are match-making festivals at spring-tide or harvest, when a great part of the year's marriages are arranged. This is not only often done among the lower races, but traces of it remain in Greece, where the dances at Megara on Easter Tuesday are renowned for wife-choosing, and till lately in Brittany, where on Michaelmas Day the girls sate in a row decked in all their finery on the bridge of Penzé, near Morlaix. The custom of bride-capture, where the bridegroom and his friends make show of carrying off the bride by violence, is known in Europe as a relic of antiquity, as in ancient Rome, Wales within the last century or two, or Tyrol at the present day; but in more barbaric regions, as on the Malay peninsula or among the Kalmuks of North Asia, it may be often met with, practised as a ceremony, or even done in earnest. On the other hand, restrictions on marriage between kinsfolk or clansfolk are more prominent among the lower races than in the civilised world, but their motive is even now imperfectly understood. Partly these restrictions take the form we are accustomed to of prohibiting marriage between relatives more or less near in our sense, but among nations at a lower level they are apt to involve also what is called exogamy or "marrying-out." A tribe or people—for instance, the Kamilaroi of Australia, or the Iroquois of North America—is divided into hereditary clans, members of which may not marry in their own clan. In various parts of the world these clans are named from some animal, plant, or other object, and anthropologists often call such names "totems," this word being taken from the native name among Algonquin tribes of North America. For an instance of the working of this custom among the Iroquois tribes a Wolf was considered brother to a Wolf of any other tribe, and might not marry a Wolf girl, who was considered as his sister, but he might marry a Deer or a Heron. In contrast with such rules is the practice of endogamy, or "marrying-in," as among some Arab tribes, who habitually marry cousins. But it will be found that the two rules often go together, as where a Hindu must practically marry within his own caste, but at the same time is prohibited from marrying in his own gotra or clan. Researches into totem-laws are apt to bring the traveller into contact with other relics of the ancient social institutions in which these laws are rooted, especially the practice of reckoning descent not on the father's side, as with us, but on the mother's side, after the manner of the Lycians, whose custom seemed extraordinary to the Greeks in the time of Herodotus, but may be still seen in existence among native tribes of America or in the Malay islands. Even the system of relationship familiar to Europeans is far different from those of regions where forms of the "classificatory system" prevail, in which father's brothers and mother's sisters are called fathers and mothers. In inquiring into native laws of marriage and descent, precautions must be taken to ensure accuracy, and especially such ambiguous English words as "uncle" or "cousin" should be kept clear of.

Another point on which travellers have great opportunity of seeing with their own eyes the working of primitive society is the holding and inheritance of property, especially land. Notions derived from our modern law of landlord and tenant give place in the traveller's mind to older conceptions, among which individual property in land is hardly found. In rude society it is very generally the tribe which owns a district as common land, where all may hunt and pasture and cut fire-wood; while, when a family have built a hut, and tilled a patch of land round it, this is held in common by the family while they live there, but falls back into tribe-land if they cease to occupy it. This is further organised in what are now often called "village communities," which may be seen in operation in Russia and India, where the village fields are portioned out among the villagers. Those who have seen them can understand the many traces in England of the former prevalence of this system in "common fields," etc. There is the more practical interest in studying the working of this old-world system from the light it throws on projects of communistic division of land, which in such villages may be studied, and its merits and defects balanced. On the one hand it assures a maintenance for all, while on the other it limits the population of a district, the more so from the obstinate resistance which the counsel of "old men" who manage a village always oppose to any improved method of tillage. Not less perfectly do the tenures existing in many countries show the various stages of landholding which arise out of military conquest. The absolute ownership of all the land by a barbaric chief or king, which may be seen in such a country as Dahome, whose subjects hold their lands on royal sufferance, is an extreme case. In the East, feudal tenures of land granted for military service still have much the same results as in mediæval Europe.

At low levels of civilisation the first dawning of criminal law may be seen in the rule of vengeance or retaliation. The person aggrieved, or his kinsfolk if he has been killed, are at once judges and executioners, and the vengeance they inflict stands in some reasonable relation to the offence committed. Not only is such vengeance the great means of keeping order among such rude tribes as the Australians, but even among half-civilised nations like Abyssinians and Afghans the primitive law may still be studied in force, carried out in strict legal order as a *lex talionis*, not degraded to mere illegal survival in outlying districts like the "vendetta" of modern Europe, carried on even now, in spite of criminal jurisprudence, which for ages has striven to transfer punishment from private hands to the State. Whether among savages, barbarians, or the lower civilised nations, the traveller will find everywhere matter of interesting observation in the law and its administration. The law may be still in the state of unwritten custom, and the senate or council of old men may be the judges, or the power at once of lawgiver and judge may have passed into the hands of the chief, who, as among the modern Kaffirs, may make a handsome revenue by the cattle given him as fees by both sides, a fact interesting as illustrating the times when an European judge took gifts as a matter of course. Among the nations at higher levels of culture in the East, for instance, most of the stages may still be seen through which the administration of law, criminal and civil, was given over to a trained legal class. One important stage in history is marked by religion taking to itself legal control over the conduct of a nation. The working of this is seen among Oriental nations, whether Mohammedan, Brahman, or Buddhist, whose codes of law are of an ecclesiastical type, and the lawyers theologians. There is much to be learnt from the manner in which such law is administered, and the devices are interesting by which codes framed under past conditions of society are practically accommodated to a new order of things, without professedly violating laws held to be sacred, and therefore unchangeable. Ordeals, which have now disappeared from legal procedure among European nations, are often to be met with elsewhere. Thus in Arabia the ordeal by touching or licking hot iron is still known (the latter is an easy and harmless trick, if the iron is quite white-hot). In Burma, under native rule, the ancient trial of witches by "swimming" went on till lately. In many countries also symbolic oaths invoking evils on the perjurer are to be met with, as when the Ostyaks in Siberia swear in court by laying their hand on a bear's head, meaning that a bear will kill them if they lie. It shows the carelessness with which Europeans are apt to regard the customs of other nations, that in English courts a Chinese is called upon to swear by breaking a saucer, under the entirely erroneous belief that this symbolic curse is a Chinese judicial oath.

The most undeveloped forms of government are only to be met with in a few outlying regions, as among some of the lower Esquimaux or Rocky Mountain tribes, where life goes on with hardly any rule beyond such control as the strong man may have over his own household. Much oftener travellers have opportunity of studying, in a more or less crude state, the types of government which prevail in higher culture. It is of especial interest to see men of the whole tribe gathered in assembly (the primitive *agora*) to decide some question of war or migration. Not less instructive are the proceedings of the council of old men (the primitive *senate*), who, among American tribes or the hill tribes of India, transact the business of the tribe; they are represented at a later social stage by the village-elders of the Hindus or the Russians. Among the problems which present themselves among nations below the civilised level is that of the working of the patriarchal system, still prevailing among such tribes as the Bedaween, while often the balance of power is seen adjusting itself between the patriarchal heads of families and the leaders who obtain authority by success in war. The struggle between the hereditary chief or king and the military despot, who not only usurps his place but seeks to establish hereditary monarchy in his own line, is one met with from low to high levels of national life. The traveller's attention may be called to the social forces which do their work independently of men in authority, and make society possible, even when there is little visible authority at all. The machinery of government described in books is often much less really powerful than public opinion, which controls men's conduct in ways which are so much less conspicuous that they have hardly yet been investigated with the care they deserve.

Religion and Mythology.—While great religions, like Mohammedanism and Buddhism, have been so carefully examined that European students often know more about their sacred books than the believers themselves, yet the general investigation of the religions of the world is very imperfect, and every effort should be made to save the details from being lost as one tribe after another disappears, or passes into a new belief. Missionaries have done much in recording particulars of native religions, and some have had the skill to describe them scientifically; but the point of view of the missionary engaged in conversion to another faith is unfavourable for seeing the reasons of the beliefs and practices he is striving to upset. The object of the anthropologist is neither to attack nor defend the doctrines of the religion he is examining, but to trace their rational origin and development. It is not only among the rudest tribes that religious ideas which seem of a primitive order may be met with, but these hold their place also among the higher nations who profess a "book-religion." Thus the English or German peasant retains many ideas belonging to the ancestral religion of Thor and Woden, and the modern Burmese, though a Buddhist, carries on much of the old worship of the spirits of the house and the forest, which belong to a far earlier religious stratum than Buddhism. It is in many districts possible for the traveller to obtain at first hand interesting information as to the philosophical ideas which underlie all religions. All over the world, people may be met with whose conception of soul or spirit is that belonging to primitive animism, namely, that the life or soul of men, beasts, or things, resides in the phantoms of them seen in dreams and visions. A traveller in British Guiana had serious trouble with one of his Arawaks, who, having dreamt that another had spoken impudently to him, on waking up went quite naturally to his master to get the offender punished. So it is reported that our officials in Burma have considered themselves disrespectfully treated when the wife or servant of the person they have come to see has refused to wake him, the

Englishman not understanding that these people hold early animistic ideas, believing the soul to be away from the sleeper's body in a dream, so that it might not find its way back if he were disturbed. As scientific ideas of the nature of life and dreams are rapidly destroying these primitive conceptions, it is desirable to collect all information about them for its important bearing on the history of philosophy and religion. The same may be said as to the ancient theory of diseases as caused by demons, and the expulsion and exorcism of them as a means of cure, which may still be studied everywhere outside the scientific nations. Information as to religious rites is of course valuable, even when the foreign observer does not understand them, but if possible their exact meaning should be made out by some one acquainted with the language, otherwise acts may be confused which have really different senses, as where a morsel of food offered as a pious offering to an ancestral ghost may be taken for a sacrifice to appease an angry wood-demon. A people's idea as to the meaning of their own rites may often be very wrong, but it is always worth while to hear what they think of the purpose of their prayers, sacrifices, purifications, fasts, feasts, and other religious ordinances, which even among savage tribes have been long since stereotyped into traditional systems.

Mythology is intimately mixed up with religion, which not only ascribes the events of the world to the action of spirits, demons, or gods, but everywhere individualises many of these beings under personal names, and receives as sacred tradition wonder-tales about them. Thus, to understand the religion of some tribes, we have not only to consider the rude philosophy under which such objects as heaven and earth or sun and moon are regarded as personal beings, whose souls (so to speak) are the heaven-god and earth-god, the sun-god and moon-god; but we have to go on further and collect the religious myths which have grown on to these superhuman beings. The tales which such a people tell of their origin and past history may to some extent include traditions of real events, but mostly they consist of myths, which are also worth collecting, as they often on examination disclose their origin, or part of it. This is seen, for instance, in the South Sea Island tale of the god Maui, whose death, when he plunged into the body of his great ancestress the Night, is an obvious myth of the sunset. The best advice as to native mythology is to write down all promising native stories, leaving it to future examination to decide which are worth publishing. The native names of personages occurring in such stories should be inquired into, as they sometimes carry in themselves the explanation of the story itself, like the name of Great-Woman-Night in the Polynesian myth just referred to. Riddles are sometimes interesting, as being myths with an explanation attached, like the Greek riddle of the twelve black and twelve white horses that draw the chariot of the day. It is not too much to say that everything which a people thinks worth remembering as a popular tradition, and all the more if it is fixed in rhyme or verse, is worth notice, as likely to contain something of historical value. That it may not be historically true is beside the question, for the poetic fictions of a tribe often throw more light on their history than their recollections of petty chiefs who quarrelled fifty years ago. The myths may record some old custom or keep up some old word that has died out of ordinary talk, or the very fact of their containing a story known elsewhere in the world may give a clue to forgotten intercourse by which it was learnt.

Customs.—It remains to say a few words as to the multifarious customs which will come under the traveller's observation. It does not follow that because these may be mentioned or described in books they need not be further looked into. The fact is that accurate examination in such matters is so new, that something always remains to be made out, especially as the motives of so many customs are still obscure. The practice of artificially deforming the infant's skull into a desired shape, which is not quite forgotten even in Europe, may be noticed with respect to the question whether the form to which the child's head is bulged or flattened is the exaggeration of the natural form of an admired caste or race. If not, what can, for instance, have induced two British Columbian tribes, one to flatten their foreheads and the other to mould them up to a peak? In tattooing, an even more widespread practice, it is well to ascertain whether the pattern on the skin seems to have been originally tribe-marks or other signs or records, or whether the purpose is ornament. In South-East Asia the two motives are present at once, when a man has ornamental designs and magical charm-figures together on his body. With regard to ornaments and costumes, the keeping-up of ancient patterns for ceremonial purposes often affords curious historical hints. Thus in the Eastern Archipelago, the old-fashioned garments of bark-cloth are used in mourning by people who have long discarded them in ordinary wear, and another case is found among some natives of South India, whose women, though they no longer put on an apron of leaves as their real ordinary garment, wear it over a cotton skirt on festival-days. Among the amusements of a people, songs are often interesting musically, and it is well to take them down, not only for the tunes, but also for the words, which sometimes throw light on old traditions and beliefs. Dancing varies from spontaneous expression of emotion to complex figures handed down by tradition and forming part of social and religious ceremony. The number of popular games in the world is smaller than would be supposed. When really attractive they may be adopted from one people to another till they make their way round the world. Any special variety, as of ball or draughts, should therefore be noticed, as it may furnish evidence of intercourse by which it may have come from some distant nation.

Though the subjects of anthropological interest are not even fully enumerated in the present chapter, some idea may have been given of the field of observation still open to travellers, not only in remote countries, but even in Europe. In taking notes, the explorer may be recommended not to be afraid of tedious minuteness, whereas the lively superficiality of popular books of travel makes them almost worthless for anthropology.*

In looking through the above remarks, written some years since, alteration has seemed hardly needful. The writer thinks, however, that it may be useful to call attention to the increased opportunities of travellers to study and obtain implements of the rudest and most ancient Stone Age. Up to a few years ago they could only have expected to find proof of the recent use among savages of stone hatchets, knives, arrowheads, etc., such as in Europe are relics of ancient tribes. These, indeed, have been known for more than a generation not to be the oldest relics of the kind, but have been called neolithic or of the New-Stone Age, to distinguish them from the far older and lower types of the mammoth period, called palæolithic or of the Old-Stone Age.

* More extended accounts of the departments of the Science of Man here noticed, and a list of works useful to advanced students, will be found in Tylor's 'Anthropology: an Introduction to the Study of Men and Civilisation' (Macmillan and Co., New edition, 1895).

Implements of this class, after their discovery in Europe, were soon noticed in India, and are now especially recognised as found over a great part of Africa. Of later years, in the islands of the South Pacific, stone implements of an even lower class have not only been found in the ground, but there is evidence that they had remained in use into modern times. In Tasmania it is on record from European eye-witnesses that tools made from chips of hard stone by trimming to an edge on one side, and which were grasped in the hand without any handle, were the cutting and hacking instruments of the natives into the last century, almost up to the time of their extinction. Thus apparently the oldest known phase of human life endured in this region untouched by civilisation, and travellers have the opportunity of studying its recent relics in Tasmania, while similar traces of rude Stone Age life, though not reaching up to so late a time, are making their appearance both in West Australia and New Zealand. Travellers should be careful to consider whether chipping is really artificial, and not due to natural action of water or wind-blown sand. There is no doubt that many "implements" in our museums are freaks of nature, e.g., those found in such quantities in the desert plateaux above the lower Nile.

Travellers of the present day have still opportunities of observation in the history of culture which will have disappeared in another generation. Inquiry in outlying countries should be made for the vanishing survivals of arts and customs, stories, and even languages. In Europe there is much of this kind to be met with by the inquirer, especially off the beaten track. Thus the dug-out canoe, the monoxyle of Hippokrates, need not be sought on African lakes, for it is still the fisherman's craft of Hungary and Bosnia; and in the same region the apparatus for producing the ceremonial need-fire by friction of wood, which disappeared from Scotland towards the beginning of last century, and the "whithorn" of coiled bark, the rustic musical instrument just vanishing from English peasant life, are still in ceremonial use. As for savage tribes which come within the traveller's ken, though their stone implements have been mostly superseded by the white man's cutlery, many arts of the remote past may still be seen. The yet simpler means of producing fire by drilling a stick with the hands without further mechanical adaptation may still be seen among savages who have not lost their old arts, and the twisting of thread with the hands, which preceded the use of even the spindle, is not everywhere forgotten. Though the study of the religion and folklore of the savage and barbaric world must be left to those who are residents rather than visitors, the passer-by who inquires may see primitive rites of religion or magic. Thus in many an Indian house in Arizona or New Mexico the traveller is reminded of his classic recollections when he sees the first morsel of the meal thrown into the fire as an offering to the ancestral spirits.

QUERIES OF ANTHROPOLOGY.—*By the late* SIR A. W. FRANKS, K.C.B., F.R.S., *Keeper of British and Mediæval Antiquities and Ethnography, British Museum.*

I. *Physical Character.*

Average height of men and women in each tribe.
Woolliness of hair.
Prognathism.
Strength in lifting and carrying weights, &c.
Speed in running.
Accuracy of aim.
Knowledge of numbers, weights, and measures.

II. *Mode of Subsistence.*

Whether mainly by hunting, or by pastoral or agricultural pursuits. Any instances of dwellings in caves.

Use of boats; forms of boats and of paddles; mode of paddling.

Any particular stratagems used in hunting, snares and traps; implements for hunting; use of dogs and of cross-bows, as well as bows and arrows.

Fishing; nets; fish-hooks; spears; any javelins or arrows with loose heads attached by a cord.

Modes of cooking, and implements used; any particular observances in cooking or at meals; any separation of sexes at meals. How is fire produced? and are any persons charged with the preservation of it?

Forms and construction of houses. Separation of the sexes.

Furniture of houses.

Plans of towns and fortifications.

Plants cultivated for food or manufactures; agricultural implements.

III. *Religion and Customs.*

Birth ceremonies.

What are the idols and their names? Is there any distinction between them in importance? What worship is paid to them? and what offerings are made, and on what occasions?

Are there any particular superstitions? What fetishes or amulets are used? by whom are they made? Are there any forms of divination, any use of casting lots with cowries, ordeals by poison or otherwise?

Vampire beliefs and ghost beliefs generally.

Cannibalism, and motives for the same.

Funeral rites. Belief in a future state. Deposit of objects with the dead, and whether deposited broken or whole, in or on the graves.

Are burial customs associated with belief in destiny after death?

It is important that the traveller should distinguish between genuine native traditions and those acquired through contact with civilised peoples.

Peace survivals among newer peoples.

IV. *Arts and Manufactures.*

Mode of spinning and weaving; patterns and materials employed.

Dyeing and nature of dyes.

Any mode of preparing and working leather.

Any knowledge of glass-making. If not acquainted with the manufacture of glass, do they melt down broken European glass and beads to make armlets and other ornaments?

Musical instruments: their forms, nature, and names.

Knowledge of pottery and mode of manufacture.

Use and manufacture of tobacco and other narcotics; forms of tobacco-pipes; any ceremonies connected with smoking; use of snuff; snuff-bottles.

Manufacture and trade in salt, wine, beer, or other liquors.

Knowledge of simple medical remedies, cupping, etc.

Ivory and wood-carving.

Metallurgy: working in the various metals, whether by a special class of people or tribes; implements used in smelting, etc. Where are the ores obtained?

Is there any knowledge of precious stones?

V. *Personal Ornaments, Disfigurements, etc.*

Are there any special marks made by tattooing or cicatrices to distinguish the various tribes? are they the same in both sexes? Drawings of these marks would be very desirable, distinguishing each tribe.

Are the teeth filed or knocked out? If the former, into what shapes are they filed? when is the filing effected? and is it the same for both sexes?

Is antimony used for the eyelids? and how is it applied?

Are ear-ornaments worn by either sex? are they pendent or inserted in the lobe? Are there any nose or lip ornaments?

Is the hair cut into any peculiar shape, or is its colour altered by dyeing?

Is any cap or protection worn on the penis, as by the Kafirs and other tribes?

Any peculiarities of dress for men and women? any distinction between married and unmarried?

What protection is worn in battle? What are the forms of the weapons? and is any missile weapon in use?

Is there any mutilation of the sex organs?

Are any marks used as distinctions for bravery, success in hunting or rank?

VI. *Ivory and Wood Carving.*

If elephant ivory is not of native origin, where is it obtained? Are any other materials of the same nature employed in carving, such as walrus-tusk, cachalot teeth, etc.? Are any very hard woods employed; and if so, how are they worked?

VII. *Money.*

What kinds of money are in use? Do the coins pass by weight as bullion, or have they a recognised value? Are any objects such as iron bars or tools, salt, pieces of cotton, cowries, beads, wampum, etc., employed as a means of exchange? If so employed, is there any recognised way in which their value is certified, or is their value the subject of bargain in each case?

VIII. *Miscellaneous.*

Any knowledge of the stars and constellations?

What games are in use? and how are they played?

Are any ancient stone implements found among the natives? and have they any superstitious regard for them?

Are any peculiar ornaments used in dancing?

Are there any modes of marking property?

Are wooden pillows in use? and do their forms differ according to tribes?

It may be added that the native names will in all cases be very desirable.

Anthropological Notes.

By W. L. H. Duckworth, M.D., Sc.D., M.A.

The following notes deal in the briefest possible manner with the more important observations to be made on human skulls; and it is conceivable that some such records might be obtained under circumstances precluding the observer from securing or removing the actual specimens, and even in instances where only a very brief period is available for inspecting them. It is convenient to arrange the observations under the following separate headings:—

1. The circumstances attending the death of the individual should be first investigated, and observations concerning the mode and locality of interment should be recorded (cf. Professor Tylor's Schedule, under "Funeral Rites").

2. The general condition of the specimen next demands attention. This is to some extent dependent upon circumstances referred to in the preceding section. From the texture and preservation of the actual bony substance, a rough estimate may be formed as to the lapse of time

since the death of the individual. The occurrence of the skull alone, or at a distance from the remainder of the skeleton, should be noted.

3. The greater weight and size, as well as the greater prominence of the brow-ridges and ridges at the back of the skull, serve to distinguish the male *sex*, but in many cases the determination of sex is almost impossible,

As regards *age*, skulls are conveniently described as immature, adult or senile. Immature skulls lack the full complement of teeth or of sockets for these, and a deep cleft is seen to cross the base of the skull about an inch in front of the large hole (foramen magnum) for the spinal cord. Upon the attainment of maturity this cleft is obliterated, by the fusion of its margins. In senile crania, the sutural lines on the surface are almost entirely obliterated, and the jaws are toothless and much reduced in size and prominence.

4. The skull may be deformed. It is important to attempt to distinguish *deformation* produced during life (either artificially or otherwise) from that determined by the weight of superincumbent soil after interment. Artificial deformation is manifested in most cases by flattening of the forehead, or of the back of the skull, or of both. But posthumous deformation is quite irregular, the face is often involved as well as the brain-case, there is often extreme flattening from side to side, and the bones are usually fragile and tend to fall apart.

5. *Trephine-holes*, indicative of enterprise in operative surgery, should be noted. The regularity of contour and the size of the hole often gives the clue to its real nature. It must be remembered that, in the process of exhumation, injuries closely resembling the foregoing operative wounds may be received by skulls, and therefore the circumstances of exhumation demand enquiry in this connection. Some skulls are found to have been incised or engraved with decorative patterns after death.

6. *Craniological descriptions* deal with the appearances presented by skulls in each of the five normal positions or aspects depicted in Fig. 1.

In the first view or aspect (Fig. 1, No. 1) the general form of the skull is shewn, and in proportions the cranial case may be either elongated or rotund; or again, if elongated, it may be elliptical (with no great difference in form between front and back), ovoid (when the hinder end is the broader), or rhomboid (lozenge or diamond-shaped).

In No. 2, the profile line, and its modification by prominent brow-ridges or jaws, claims attention. In No. 3, the general form of the eye (orbital aperture) and nose (nasal aperture), as well as the relative breadth of the face, are considered. No. 4 shows the palate, and the number and forms of the teeth are studied from this point of view. In No. 5, the form of the transverse cranial arc, and any irregularities, such as flattening or the production of a keeled (scaphoid) appearance, should be noted.

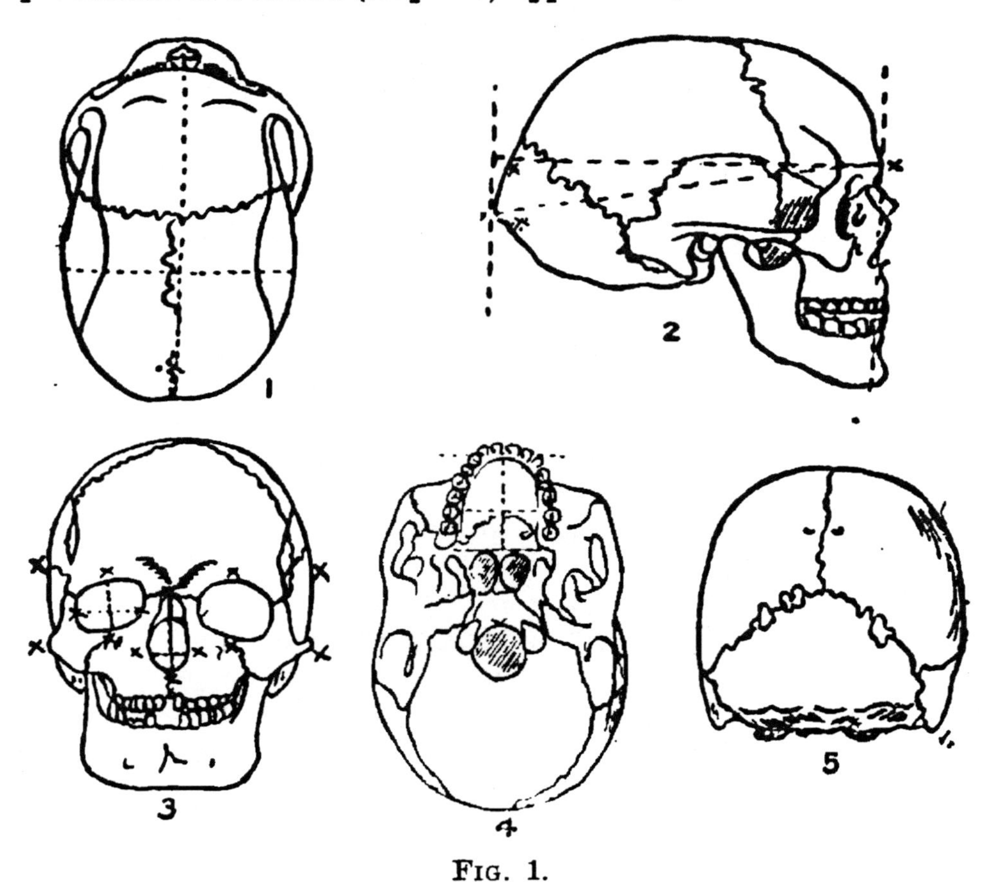

FIG. 1.

The five normæ, or aspects of the human skull, viz.:—(1) Norma verticalis, the vertex view; (2) Norma lateralis; (3) Norma facialis; (4) Norma basilaris; (5) Norma occipitalis.

7. The *lower jaw* is studied independently. The prominence of the chin, the squareness of the angle, the stoutness of the whole bone, and the number and characters of the teeth are the chief points to which attention is directed.

8. *Cranial measurements* are made with callipers and a flexible steel or tape measure. These objects, as well as a graduated two-meter rod used for measuring stature or long bones, are supplied in a travelling case by Messrs. Hermann of Zürich, who have made them to Professor Martin's designs. The whole outfit costs about £4, but the instruments can be obtained separately. Messrs. Hermann also make a modified

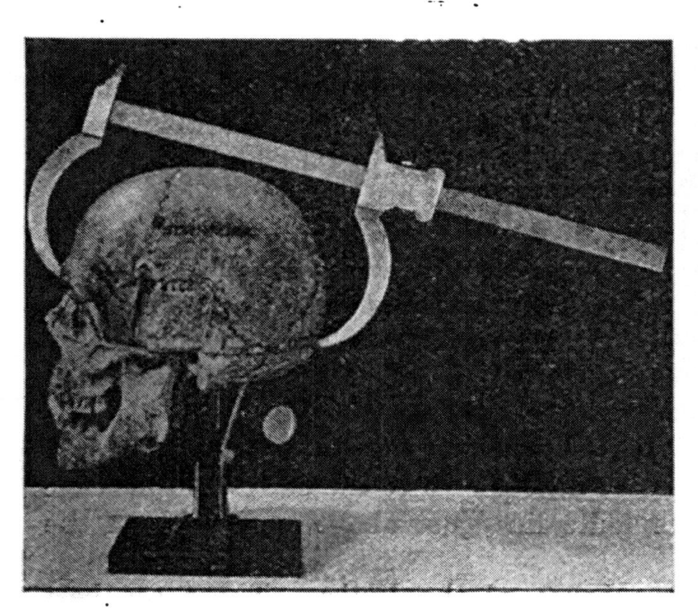

FIG. 2.—Measurement of the length of the skull with callipers (Flower's Craniometer as modified by Dr. Duckworth).

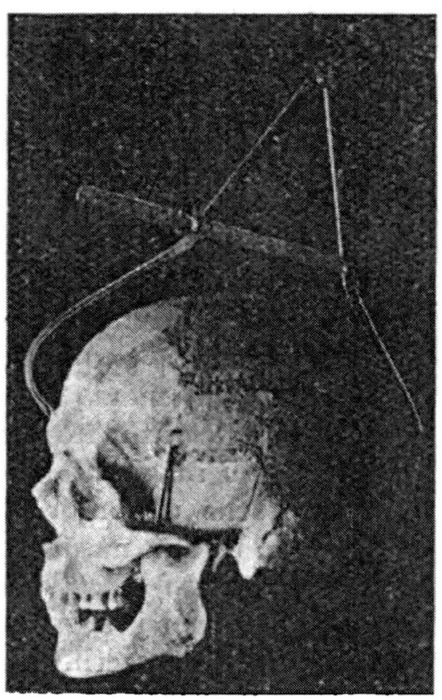

FIG. 3.—Measurement of the length of the skull with another form of callipers.

form of the callipers, known as Flower's Craniometer (cf. Fig. 2). Measurements are recorded in millimeters, in which the various callipers, etc., are graduated. The latter instruments may have the forms represented in Fig. 2 (Flower's Craniometer) or Fig. 3, in which are shown the exact positions of the instruments in measuring the length of the skull.

The chief measurements may be now enumerated in order of importance:—

i. The *extreme length* of the brain-case, measured as shown in Figs. 2 and 3. In Fig. 1 (No. 2), the lower of the two dotted lines traversing the skull shows the approximate direction of the measurement, which, it may be repeated, is the extreme or maximum length (in the middle line) of that part of the skull situated above and behind the face.

ii. The *maximum breadth* of the skull is measured on the brain-case, as shown by the transverse dotted line in Fig. 1 (No. 1).

iii. The *circumference* of the brain-case (as distinct from the face) is measured with the flexible tape passed round the brow-ridges and the back of the skull, as is indicated by the lower dotted line in Fig. 1 (No. 2). The measurement is that of the greatest circumference of the brain-case obtainable in this way.

iv. Measurements illustrative of the degree of prominence of the upper jaw are two in number, and are made with the callipers. The two measurements start from the same point behind; this point is the middle of the front edge, or margin, of the hole for the spinal cord (foramen magnum), shown in Fig. 1 (No. 4). From this common point, the upper or *basi-nasal measurement* passes to the upper border of the nasal bones in the middle line of the face, and the lower or *basi-alveolar measurement* passes to the extreme front edge of the upper jaw in the middle line, just above and between the upper incisor teeth.

9. *Measurements of the long bones* of the limbs. These are best made by means of a graduated rod with fixed and movable limbs, resembling a large pair of callipers. Such a rod is supplied (as indicated above) by Messrs. Hermann. The bones to be measured are six in number for each side of the body, viz.:—those of the upper arm and thigh (humerus and femur); then two for the fore-arm (called radius and ulna, the former being the shorter); and two for the leg (called tibia and fibula, the former, or shin-bone, being the larger). The extreme or maximum length is measured in each case, excepting those of the femur and tibia.

i. The femur is measured obliquely, that is, from its ball-like head to a line touching *both* prominences at the lower end, as shown at B in Fig. 4. This represents the position of the bone in a person standing erect.

ii. The spiny processes on the upper end of the tibia are not included in the measurement of its length.

Immature limb-bones are recognised by the line which demarcates the extreme ends, or epiphyses, from the shaft (or diaphysis). Sexual differences are chiefly indicated by greater length, stoutness, and also by the development of osseous ridges in male bones.

FIG. 4.—Position of the thigh-bone for measurement of its length.

10. The measurements of the long bones are used in estimating the *stature.* Special notice is to be paid to individuals of giant or of dwarf stature. In this connection particular mention should be made in the following instances:—

(*a*) Where the thigh bone measures more than 519 mm., or less than 363 mm.

(*b*) Where the shin bone measures more than 420 mm., or less than 284 mm.

(*c*) Where the fibula measures more than 413 mm., or less than 283 mm.

(*d*) Where the upper arm bone measures more than 368 mm., or less than 263 mm.

(*e*) Where the ulna measures more than 293 mm., or less than 203 mm.

(*f*) Where the radius measures more than 273 mm., or less than 193 mm. (only adult bones being considered).

11. Adult skulls will also excite attention when the horizontal circumference (measured in accordance with the directions provided above) is greater than 556 mm., or less than 468 mm.

VI.

INDUSTRY AND COMMERCE.

By J. S. Keltie, LL.D., *Secretary, R.G.S.*

The kind of information desired under this heading may be summed up in the three following questions, so far as uncivilised or semi-civilised countries are concerned :—

(1.) What are the available resources of the country that may be turned to industrial or commercial account?

(2.) What commercial products can find an available market in the country?

(3.) What are the facilities for or hindrances to intercourse between the country and the rest of the world?

Or, briefly, (1) Resources; (2) Wants; (3) Accessibility.

These include the questions of suitability for immigration and colonisation.

What is known as commercial geography is one of several special applications of geographical knowledge. From this practical point of view, therefore, the observations collected under other heads in this book will be of service, especially if the requirements of commerce are kept in view at the same time as the *desiderata* of science. From this standpoint, the sections on Meteorology, Geology, Natural History, and Anthropology should be consulted. Even general geographical and topographical observations will be of practical service—the general lie of the country, its altitudes, and its character at certain altitudes, its mountains, hills, valleys, plains, rivers—if regarded from the special standpoint of habitability and possibilities of development.

The suggestions contained in this, as in the other sections of this manual, are meant both for the ordinary explorer or traveller who may have to pass rapidly through a country and for those who may have more opportunity for making leisurely observations. To the former the following brief hints may prove serviceable:—

Observe and note—

1. As regards RESOURCES:—

The extent or quantity, quality, facilities for raising or collecting, for transport and shipment, etc., of

(*a*) The natural products, such as minerals and metals, especially gold, silver, coal, iron, copper, tin, nitre, guano, phosphates, etc.; timber useful for various purposes; vegetable products, useful for food, fibres, dye-stuffs, or medicine—gums, resins, etc.; animal products useful for food, skins, fur, feathers; riverine or lacustrine products, useful for food, oil, or other purposes.

(*b*) Substances cultivated for food or for manufacturing purposes.

(*c*) Articles manufactured for clothing, for domestic, warlike, or other purposes.

(*d*) Native methods of developing resources, of carrying on manufactures, and of transport.

(*e*) Practicability of developing the resources of the country by European methods.

2. As to WANTS:—

(*a*) What particular things used by the natives could be profitably supplied from the outside?

(*b*) What do the natives lack that might be introduced and supplied from the outside?

(*c*) In all cases be particular to note favourite materials, shapes, colours, or other peculiarities, as of cloths, implements, ornaments, such as beads, anklets, etc.

3. As to ACCESSIBILITY:—

Note the nearest ports and railways; the character and connections of native roads, if any; the navigability of rivers, inlets and lakes for various kinds of craft at various seasons of the year; heights of passes, whether suitable for animals or only for porters, whether blocked at any season of year; routes likely to be suitable for railways.

In all cases where practicable, specimens of products should be obtained, in order that specialists at home may judge of their industrial or commercial value.

For those who may have more leisure for observation, some or all of the following detailed hints may be useful:—

MINERALS AND METALS.—It is not necessary to add much here to what has been said under Geology. If the traveller is not himself competent, or has no opportunity to test the value of these products, he should bring home specimens; this, if possible, should be done in any case. Under this section a look-out should be kept for any indication of naphtha, asphalt, or mineral oils. From the commercial point of view the important points are—

Quality.—To what extent are minerals or metals, as gold, copper, iron, phosphates, mixed up with other matter? What is the yield per ton of ore? In the case of coal, how does it burn, and what is the percentage of ash? Next—

Quantity.—Does the substance occur in sufficient quantity to make it worth expenditure of capital and labour? The information must be obtained by personal inspection. Finally—

Locality.—Is the situation of the deposits easily accessible? How are they situated with reference to routes, existing or practicable? and how with reference to ports of embarkation? Could they be worked with the resources available in the country, or would labour and machinery require to be introduced? If worked in the country, is there any neighbouring market for the manufactured products? What are the native processes (if any) of obtaining and working minerals?

We must again refer to the section on Geology for further details, and the intending traveller would do well to take a few lessons before he leaves, so as to be able to recognise the most common and useful minerals and the conditions under which they usually occur; he will thus save much time and trouble.

Vegetable Products.—The directions for observation and collection given under the Botanical section should be attended to; and it is important that the traveller should be able to recognise the chief classes of plants, so that specialists may be able to pronounce generally on their utility.

Character of Surface.—The general character of the surface of a country, so far as its vegetation is concerned, should be clearly grasped. What proportion, or, if possible, what area is under forest? what under grass? what desert, or mountain, or marsh, or uncultivable? what under cultivation?

Forests.—If of a generally forest or thickly-wooded character, are the forests extensive and dense, with much undergrowth, as in tropical South America? or easily penetrable, as the forests of Europe and North America? Or are the trees scattered, either in clumps, or singly, as in a great part of Central Africa? Do they prevail over the country generally, or are the river-banks only lined with dense tree vegetation? Are the forests only found in the low country, or do they cover the hills and extend up the mountains?

Timber.—Ascertain the leading characteristics of the trees of the forests. What are the prevailing families, and, if possible, genera and species? What uses, if any, do the natives make of the woods? Which do they use for their houses, their furniture, their canoes, their weapons, their ornaments? How do the woods seem to stand tear and wear, the climate, the attack of destructive insects, immersion in water? Are there any woods that would do for such purposes as railway-sleepers or telegraph-poles in the conditions which prevail in the country? Any ornamental woods suitable for cabinet purposes?

Fibres, Fruits, Chemicals, etc.—Are there any plants the fibres of which could be turned to account? Any fruits adapted to human consumption, and are they found in any quantity, or could they be cultivated?

Are there any trees or other plants suitable for drugs or chemicals—bark, leaves, juices, roots? What medicines, narcotics, or stimulants are used by the natives? how are they obtained and how prepared?

Are there any species of useful plants growing wild—coffee, sugar, cotton, vanilla, spices, etc.? Any trees producing gums that might be of commercial value, like gum arabic, gutta-percha, or caoutchouc? Or any whose fruits yield oil, like the cocoa-nut and the olive? Do the natives make use of these juices? What are the processes of extraction and preparation?

Other Vegetation.—When there is an undergrowth, its character should be noticed, and the diseases, if any, to which trees are subject.

Note what other vegetation exists besides that of trees. Are there any plants like the turnip, the potato, the batata, which are useful as foods, or for other purposes? Specimens of any herbs likely to be useful should be obtained, especially if they are used by the natives for medicine, for dyeing, for poison, or other useful purposes.

Sometimes, as in Central and Western Australia, what arboraceous vegetation exists, consists mainly of shrubs, the character of which should be noted. Do they hinder locomotion? Are their shoots useful for forage? Are they injurious to horses and cattle?

When there is herbaceous vegetation of any extent, what is its character? Is it tall and coarse and reedy, like much of the African grass? Or such as is found on the prairies and pampas? Or of a troublesome spinifex character, as in Australia? Or of a turf-like character, like the grasses of Europe? What are the components of this kind of vegetation, and how far is it likely to prove useful as fodder? What uses do the natives put it to, either for their animals or for manufactures? Do they use it for making mats or cloth? Are there any plants mixed with it injurious to animals? What is the condition of the grass at different seasons of the year? When is it at its best, its strongest, its densest? Is it liable to be parched up at any season? To what extent is its condition affected by the climate, by rainfall, by irrigation, natural or artificial? Is it easily removed, in order to make way for other cultures? Does it spread into the forest region, and has it any special characteristics there? How does it, as well as other useful vegetation, vary with altitude or other local conditions? Do the products change under different agricultural conditions? e.g., some are poisonous under certain conditions, not others?

Marshes, Deserts, Irrigation.—Note if marshes or peat-bogs, or other special features of the surface exist to any extent, and whether the drainage of marshes is practicable.

Where deserts exist, note their character. Are they sandy, gravelly, rocky, salt? What is the prevailing rock? If the desert character of the land (as is generally the case) seems due to want of water, is there any artificial means likely to be available for supplying that want? Is there any storage of water and irrigation among the natives? and, if so, how is it accomplished? and what are the results? Are there any sources within reach, either above or underground, from whence a supply of water for irrigation purposes could be obtained? Indicate any exceptional defects of quality in the supply of water.

Note if any part of the country is liable to periodical inundations. At what periods of the year do they occur? Are these inundations destructive, or are they utilised for agricultural purposes? Would it be possible to regulate these inundations?

AGRICULTURE.—The general outcome of all these observations is the suitability of a country for agricultural development. What articles do the natives cultivate, if any? Has the cultivated land any special character, or is it simply the ordinary land cleared of trees or grass, or other wild growth? Note the methods and implements of culture used by the natives; the seasons of sowing and reaping, and preparing the crops for use. Do they depend for water on rainfall or irrigation? To what altitudes is cultivation carried, and what are the crops that prosper at these altitudes?

From the point of view of colonisation and agriculture, precise information as to the nature of the soil is desirable. The proportion or extent of a country suitable for agriculture might be noted. Observe, as precisely as practicable, the nature and depth of the upper layer of soil. The depth—it may be a few inches, or it may be two or three feet—can easily be ascertained. A general idea may also be given of its nature. Is it mostly vegetable mould, as it is likely to be in old forest or grass countries? or peaty? or marshy? If possible, also, ascertain the depth of the subsoil down to the rock or clay, or other permanent basis on which it rests. In a general way it might be observed whether the soil is sandy, gravelly, stony, calcareous, marly, clayey. Also is it compact, tenacious, or loose, and, above all, is it permeable or impermeable to water? Is the soil very dry or very moist? or what is its intermediate stage?

If the natives carry on cultivation, ascertain, if possible, the yield per acre of what they cultivate. Do they cultivate only for their own wants? If not, where and what is the nature of the market to which they send the surplus?

Does the country seem suitable for other cultures besides those carried on by the natives?

Animal Products.—If there are wild animals in the country, observe whether the natives hunt them for what they yield in the way of food or other useful products. Are there any ivory-yielding animals, or animals whose skins can be turned to profitable account? Could a sufficient supply for mercantile purposes be obtained by means of native hunters, if properly encouraged; or how would it be best to work such resources? Are there any laws or customs enforced by the natives in hunting wild animals? Are there any noxious wild animals, and to what extent do they affect human comfort and human life?

Domestic Animals.—It is important to know what domestic animals the natives possess, how they are reared and fed, and what uses they are put to. Also whether the country is good for horses, cattle, sheep, and poultry, and approximately what is the extent, situation, and accessibility of the grazing-lands. Are the pastures perennial? To what extent do they depend on rainfall, or irrigation, or on intermittent streams? During what months of the year are they available? Are there any plants among the pastures injurious to animals? Are there any insects (like the *tsetse*) or other animals injurious to cattle or horses? Do horses or asses exist among the natives, and what uses are they put to? If not, would they be likely to flourish, if introduced?

Fisheries.—Information concerning fish and fisheries is desirable; and among fish, from the practical standpoint, are included shell-fish (especially pearl shells), sponges, corals, and animals of the whale and seal kind. If the natives practise fishing, either in lakes, rivers, or the sea, ascertain the kinds of fish they capture, their methods and implements, and the particular seasons at which fishing is practised. Are the fisheries, whether worked by the natives or not, likely to be of commercial value?

Trade.—Much of the information suggested above will be of service from the special commercial point of view, especially with reference to export. Information should be obtained concerning any manufactures carried on by the natives besides what has been suggested above—manufactures in metal, in wood, in clay, or stone; in materials derived from the vegetable and animal kingdoms, what they are, what uses they are put to, what processes are used, and to what extent, if any, they form articles of trade. With regard to the import market, what generally are the wants of the natives and what new wants might be created. If possible, some approximate estimate of the value of the leading classes of imports, if there are any, should be obtained.

Ascertain if any goods are brought into the country from the outside; if so, what they are, where they come from, and as accurate an estimate as possible of quantity, or value, or both. In the case of imports from civilised countries, are those of any particular country preferred, and, if so, why? Is it owing to anything special in quality, or pattern, or cost, means of communication, or in quantity available? Is there any special tribe of middlemen who prevent the inland people from coming into direct relation with traders? Probably a market could be created for outside manufactures which have not as yet been introduced into the country; or such manufactures might easily obtain a market in preference to those of native make. Note especially the patterns of articles of native make, as these are probably adapted to the conditions of the country, and should therefore be imitated or improved upon in the case of imported goods, the quality of the latter being better, and the cost, if possible, lower. British traders, especially, often incur serious losses by not ascertaining in advance what particular articles and what special patterns are in demand among natives. Note also whether European methods might not be introduced with advantage for the manufacture of native goods. Native usages as to credit ought to be ascertained, what safeguards are binding, what is the medium of exchange, etc. It would be useful to discover beforehand whether the importation of machinery and explosives is likely to be unimpeded.

CLIMATE.—This is an essential item, so far as the exploitation of tropical and semi-tropical countries by Europeans is concerned. Its main elements are determined by temperature, latitude, altitude, and rainfall; the character of the surface should also be taken into account. Under Meteorology, the main directions on the subject are given. The temperature at different seasons and at different times of the day (say 9 A.M., 3 P.M., and 9 P.M.) should be ascertained, and that at various altitudes. Rainfall observations are not of much service unless they can be obtained over a continuous series of years. Ascertain the distribution of rainfall over the year, and the limits of the rainy period of the year, when such period exists, and, if possible, the quantity which falls in the different months of the period; how does the rainfall differ with altitude and other topographical variations? If a country is subject to droughts, it is important to ascertain if there is any periodicity in these droughts, and how they affect the resources and prospects of the country, and the condition of the rivers. Would it be possible, by storage of water or other means, to counteract to any extent the bad effects of drought?

What effect has the rainy period and the drought period on the native inhabitants, and especially on Europeans? As far as possible, ascertain the birth and death rate per annum.

FACILITIES AND HINDRANCES TO COMMERCIAL DEVELOPMENT.—Under this head the first consideration is Accessibility or Means of Communication. Generally, the quickest, cheapest, and safest routes to a satisfactory market should be ascertained.

Water Communication.—Observe what natural means of communication exist, what is the nature of communication between the country and the outside world. If communication by sea is of importance, how is the interior to be reached from the sea? Are there any deep inlets? Is there practicable river communication? For what sort of vessels is it adapted? Are there any obstructions in the river, and, if so, what is their nature, and how could they be surmounted? What is the width of the river, the depth of the water, and the force of the current, at various distances from the mouth, and at the periods of the year when the river is at its lowest and highest? Are there any lakes that could be utilised for communication?

Roads.—If there are native roads, state precisely what is their nature under various conditions of weather; their width; what sort of vehicles, if any, they are suited for, and where they lead to. If the country is mountainous, ascertain the principal passes, what places they connect, their exact heights at the highest point (not the heights of the mountains), for what animals or vehicles they are practicable, or if only for porters, and what is their condition at various seasons of the year. If the natives have any vehicles, or vessels, or other means of transport, describe them.

Railway Routes.—Observe, as far as possible, suitable routes for railway or canal communication, as well as for good roads, and whether any materials for railway construction are obtainable in the country.

Possibly the country, if an inland one, could be connected by road or rail with some existing railway system. What are the nearest telegraph stations?

Labour.—Another important consideration under this head is that of labour—labour for the varied enterprises connected with the development of a country's resources. Is labour obtainable within the country itself? If so, to what extent, and on what terms? Are the natives industrious, and are they likely to labour under foreign superintendence? Does slavery or forced labour in any form exist? If not obtainable in the country, how may it be most easily and cheaply obtained? Is the country suited to manual labour by whites? If there are only certain kinds of labour in which whites may engage, state what they are.

To what extent could machinery be used with advantage? Is there any water power available? or any animal power?

Currency, Tariffs, etc.—Under this head also questions relating to currency should be included, or whatever other medium of exchange exists.

Another important consideration here is the question of tariffs, which, in one shape or another, exist in nearly all uncivilised and semi-civilised countries, from the hongo of Central Africa to the Customs duties of Eastern countries. Precise information concerning these, both as to exports and imports, is important.

Note, also, as precisely as practicable, the cost of living for Europeans settling down for a time, and the expenses involved in travelling through a country.

Inhabitants.—Is there anything in the character of the natives—physical, mental, or moral—likely to affect commercial intercourse or the industrial development of the country? Any prejudices or superstitions that should be attended to? Anything in the attitude of natives to traders and settlers deserving consideration? Is the population nomadic or settled? What material, if any, do they use for smoking, and what is the nature of their intoxicating drinks, if they have any? Estimate, as nearly as possible, the population, the density per square mile, both for the country as a whole and for the chief centres of population. Ascertain the nature of any political or social organisation which exists. What are the terms on which land can be acquired? What are the prevalent crimes? Under the Anthropological section directions are given for ascertaining the leading racial characteristics of the people.

VII.

ARCHÆOLOGY.

By D. G. HOGARTH, M.A.

THIS section is intended, not for experts, but for those travellers without special interest or experience in archæological work, who, finding themselves in districts where unpublished antiquities exist, feel it incumbent on them to record or rescue them. General hints, therefore, are subjoined on methods of recording, cleaning, temporarily conserving, and conveying monuments and objects of antiquity.

A.—RECORDING.

There are three ways of obtaining a record of a monument, all of which should be used if possible in cases of importance :—

1. Mechanical reproduction by photography, moulding, or rubbing.
2. Reproduction by planning, drawing, or otherwise copying.
3. Literary memoranda of characteristics, *e.g.*, dimensions, subject, etc.

1. *Mechanical Reproduction.*

(*a*) *Photography.*—This subject has already been dealt with in Part II., and it is only necessary here to add a few hints, peculiarly applicable to the photography of objects of antiquity.

Of inscriptions photography does not, as a rule, supply a useful reproduction, and unless time or other opportunity be wanting, it should not be the only mechanical process applied. If an inscription is at all worn, or the material on which it is engraved contains superficial faults, do not rely on a photograph ; but if one is taken, try to dispose the object so that light falls on it from one side, with a slight obliquity from above. In the case of rock inscriptions or others still in their place on walls, etc., a photograph should be secured to show relative position and surroundings. An inscription on dark material will often need preparation with white chalk before being photographed to any good purpose; but the traveller who has opportunity to do that, will probably have equal opportunity to employ the better process of moulding.

For *sculpture*, photography will often be the only method of reproduction possible to the traveller. When that is the case he may be advised (1) to take his photographs on the largest possible scale; (2) to take the object from all possible points of view; (3) to do what he can to improve the light and relief of the subject. Small objects gain enormously by reflected lights carefully arranged to bring out their contours; if sufficient mirrors are not obtainable, strips of tin or even white paper will serve as reflectors. Backgrounds may be chalked or blacked as the case may be, but in the case of a relief this should only be done when the outlines are very clear. If a sculpture on rock or other material is much broken or worn, a good mechanical method of improving relief, which Professor Petrie recommends, can be used; this consists in dusting the face with powdery sand and then fanning it. The result, if carefully done, is to leave hollows and background in strong contrast to the surface of the relief. Highly polished or lucent surfaces should be dulled before photographing. Metal objects do not, as a rule, photograph very satisfactorily; and of coins it is much better to take a cast and photograph that. Perhaps it is not superfluous to remind the photographer to make notes of all colouring before leaving his subject. On preparing sculpture, etc., for photography there are valuable practical hints in Professor Petrie's '*Methods and Aims in Archæology*,' Chapter VII.

For *small objects of antiquity*, photography is not of much service unless a camera with very long extension can be used. If it be possible, the best way is to lay the objects on a sheet of glass raised about a foot from the ground on a frame, and arrange the camera above, so as to photograph downwards. This process obviates all ground shadows, and all pins or other supports for the objects, which appear sharply on the negative as if suspended in air.

For *buildings* photography is, of course, the only method of mechanical reproduction possible.

(*b*) *Moulding*.—This in almost all cases will have to be done by impressing *paper*. Only in the case of small objects will a traveller ordinarily be able to use either *plaster of Paris* (if he can procure it) or *sealing wax*. As to the latter he may be reminded never to heat his wax by putting it in direct contact with a flame, or the impression will come out so parti-coloured or black as to be of little use; a card should be held over a candle or lamp, or, better, a spirit burner, and the wax rubbed upon the gradually heating upper surface till enough has melted on to the card without boiling. A useful impression may be obtained of coins, gems, etc., by pressing ordinary tin foil upon the surface; but great care must be taken afterwards that the impression does not get flattened out in transport. Mr. Petrie recommends floating impressed foil on water face downwards, and dropping hot wax upon it, as a solidifying agent.

As to the use of plaster of Paris it may be remembered that (1) the proper mixture is just as much plaster as will absorb the water, leaving none standing on the top; (2) the object must be well cleaned and soaped before being moulded; (3) the plaster must be applied *very rapidly*, in a thin coating all over the object at once: it can be backed with more plaster afterwards; (4) a surface of any size should either be moulded in sections, or, if done all at once, the plaster will probably have to be cut into sections afterwards for transport. To cut it, it is best to lay strings upon the object before moulding, which, later, can be pulled up through the plaster while still viscous.

Paper moulding or '*squeezing*' is, however, the ordinary process employed. Any fairly strong *unsized* paper will serve more or less well, but a special 'squeeze' paper is procurable in most large cities where there are archæological museums (*e.g.*, in London, from Nutt's, 57 to 59, Long Acre, W.C.; in Paris, from Moreau's, 11, Passage du Pont Neuf; in Berlin, from Ebers Brothers. It can also be got direct from the maker, Papierhändler Dorr, Spiesgasse, Strassburg.) Mr. Maudslay recommended in a previous edition a hand-made paper, used for packing oranges in Spain, and to be obtained of Messrs. Batalla, of Cacagente, near Valencia, through the agency of Messrs. H. King & Co., Cornhill. Failing these, the paper on which the commoner news-sheets are printed will do. The other implements needed are a sponge, and close-bristled, not very hard, brushes of two or three sizes, *e.g.*, an ordinary clothes-brush, a nail-brush, and a tooth-brush, if no others are at hand, but special brushes with hand-straps on the back, or curved handles to keep the knuckles of the beater away from the stone, are preferable. In the case of an ordinary inscription, of not specially rough or uneven surface, brush and clean the stone thoroughly, and pick all accumulation out of the letters; then wet it thoroughly all over; lay the dry paper, cut to size, as flat on the surface as possible, and dab it down with a very wet sponge till thoroughly soaked; take the largest brush and pound the paper rather gently all over, till it partially adheres; then take smaller brushes and work the paper into all depressions with the maximum pressure you can exert; finally, hammer it with the large brush again, working systematically from top to bottom or side to side, not minding the surface becoming mashed, and driving the air-bubbles, which collect under the paper, before you and out at the farthest point. If the stone was thoroughly wetted before the paper was laid on, these ought not to be numerous or troublesome. Then, if the surface of the stone has penetrated through the paper at any point, lay a second sheet and, if necessary, a third, or more, and treat as the first was treated. Leave the sheets, if possible, to dry on the stone, and all will come off as one with a perfect reverse impression of the stone's surface. Some soak the paper before laying on the stone, but if that is done and there be any wind, the paper, become very tender, will be apt to tear with its own weight while being laid on, and will be difficult to lay flat. If for any reason, *e.g.*, overhanging of the face, it is difficult to make the paper adhere, the task will prove easier with small pieces. But whenever a surface is squeezed by sections, the operator must be careful to make the edges of his sheets overlap, so that later all can be gummed together as one sheet; to number the sheets according to a key-plan, recorded in his note-book, and to mark on the sheets themselves the lines of junction. All superfluous edges should be peeled off, as they are apt to lift in the wind and cause all or a part of the impression to become detached before it is well set, and weights should, if possible, be laid on the drying paper. If for any reason the paper cannot be left to dry in position, peel off carefully and lay out to dry reverse side uppermost. The impression, so taken, will be hardly less good. When dry, roll the squeeze inside a tin cylinder, and only very rough usage will harm it.

Squeezing sculptures is a more difficult matter. Mr. A. P. Maudslay gave very full directions in an earlier edition of these "Hints" which may be repeated here. After stating that all moulds of sculpture have to be made of many thicknesses of paper, with a good coating of paste between every few sheets (not only because of the great inequalities of surface, which break through the paper, but also in order that the mould, when dry, may be stout enough to keep its shape), he said :—

'Paper can only properly be applied for the purpose of moulding when the carving is free from large contours and deep undercutting; but it is wonderful what accurate results can be obtained even when large curves and some undercutting have to be contended against. Where worn or splintered parts of a wood-carving, or fissures in a stone, or deep undercutting which is not essential to the design, occur, it is often of advantage to fill them up with clay or paper, to which a smooth surface can be given, so that the mould will come away free from them when it is dry; and careful notes and measurements will often enable one to restore the contour to a mould which has suffered some pressure in transport. In a properly-made mould the detail of carving is never lost, unless the paper itself is destroyed.

'A shallow tin bath (or two made to fit inside one another), large enough to hold an open sheet of paper, is useful for soaking the paper in.* Twenty sheets or more may be placed in the water at once, and left there without harm for an hour or more; but a few minutes' soaking is quite enough.

'I have several times had to mould, in America, the whole of a monoithic monument—one as much as 25 feet in height—covered with carving and hieroglyphic inscription, and have been perfectly successful in reproducing it in plaster in England. Each face would be marked out into three or more sections, and each section would be moulded separately, great care being taken that each mould should considerably overlap the margin of the other, so that when each section is cast in plaster the edges of the cast can be cut away until the joint is perfect. And each section should also overlap at the top and sides in No. 1, and at the sides in No. 2, &c., for the same reason; and it is necessary to pay careful attention to the beating in of the paper near the sides and edges, as it is there that the layers are most likely to come apart when dry. These edges can be trimmed down afterwards, if found too bulky in packing.

---

* In the case of sculptures, owing to their deep depressions, it is best to wet the paper *before* laying on.—*Ed.*

'The first sheets of paper should always be put on singly, and well beaten in. If the carving presents many sharp angles, the paper will again and again be broken away over them, and small scraps of paper may be used for covering them up, until the whole section is covered at least three papers deep in the thinnest place. The coat of paste should then be given. If the paste is laid on when the mould is too thin, it will penetrate to the stone, and prevent the mould coming off when dry. The paste may be put on warm, but if too hot it draws the paper from the stone (if it is a stone sculpture), air gets underneath the paper, and it is very difficult to get rid of it again. Avoid, in putting on the first papers, doubled edges or creases, and beat in well, so that the paper may work into the grain of the stone or wood. It is easy to spoil a mould by scamping the work in it, but not easy to spoil it by overbeating. After giving a coat of paste with a brush, it is advisable to work in the paste with the fingers, so as to be sure, from the smooth feeling, that it penetrates the paper over the whole surface.

'After the first coat of paste has been applied, a good deal of time may be saved by employing an assistant to beat out the paper for the further thickening of the mould, for when thus beaten out, two or three thicknesses of paper can be laid on at the same time. Take about six sheets together from the water, fold them, and then double them twice, and slightly tear the wet doubled edges, so that when the sheets are laid open again there are a number of small slits in the paper; then lay them out together on any flat surface, and beat them out with a brush for a few minutes. It is easy to separate them again into the required number of sheets in thickness.

'Another method which is equally good, if not better than the last, is, after making the tears in the doubled sheets as before, to unfold them, and then to roll them together and twist them up like a rope, and rub them well between the hands; then unroll them and beat them out for a moment, separating as many sheets as are required. Either of these processes loosens the fibre of the paper, whilst the slits prevent it stretching unevenly. After this treatment it feels to the touch more like wet leather than paper.

'As the mould grows thicker the pulpy paper will, from the continual beating, find its way into, and fill up, the deeper cutting; but it should be most carefully watched that the mould is not left too thin over the more prominent parts of the surface, and, with a little practice, the thickness is easily judged by the touch. It is always well to use the fingers frequently both in pressing the paper into its place and working in the paste.

'It is difficult to lay down any rule as to the thickness of a mould and the number of coatings of paste necessary. If the mould is of large size, and the carving presents prominent angles or large curves, it may need an average of thirty sheets in thickness to preserve its shape, and three or four coatings of paste; but if the carving is in low-relief on a flat surface, less than half the thickness will suffice.

'In hot weather, out of doors, a mould will take about twenty-four hours to dry; but it should be covered up at night from the dew. In damp forests or in bad weather I have dried most of my moulds by building up large wood fires at the distance of a few feet from the sculpture.

'It is best to take off a mould when it is cool—in the morning or evening. *Don't be in a hurry about it.*

'If the mould is torn or broken in taking it off the carving, mend it with paste *at once.*

'When a mould is taken off, lay it to dry in the sun on a flat surface, as there is usually some moisture left in it. If the mould is not flat in shape, support it carefully, so as to preserve the contours.

'When the mould is quite dry, it is advisable, but not necessary, to give it, both back and front, one or more coats of boiled linseed oil. Heat this oil before applying it, and it will then soak in well, and use rather a soft brush, and be careful, in oiling the surface of the mould, not to rub too hard. As the paper easily absorbs moisture, the moulds need to be carefully packed.' Plaster casts can be taken very successfully from paper moulds. See directions for casting above.

(*c*) *Rubbing* can only be practised on a fairly smooth surface, and has nothing to recommend it except the ease and celerity with which it can be done. The traveller may as well carry a little heel-ball in his kit. Any thin, tough paper will do.

2. *Other Reproductions.*

Planning and *drawing* are dealt with in Vol. I. of these 'Hints.' *Copying*, as distinct from either, applies to inscriptions. A hand-copy of an inscription as well as a mechanical reproduction should always be made, partly because it has the better chance of surviving the accidents of travel, partly because, if a stone surface is at all perished, anyone with a keen eye and power of concentration ought to see more lettering on its worn parts than will appear on a photograph or 'squeeze.' The copy should be as near a facsimile on a reduced scale as time and other considerations will allow. It should be made on ruled paper, if possible on paper ruled in squares (*en quadrille*), and the relative position of the letters to one another must be kept as far as possible. All broken parts of letters are to be scrupulously copied (they can often be distinguished from stone-flaws by feeling along the bottom of the groove with a knife-blade; if that finds an even line, the groove is probably part of a letter), and all intervals where letters have perished beyond the copyist's power to recover them should be measured, and by comparison of equal intervals containing decipherable letters in other parts of the inscription, the number of lost letters can be estimated. These should be indicated by dots in a shaded patch. Letters, about which the copyist is not quite certain, must be dotted or drawn in faint line. If a stone be imperfect at the sides, and there be any means of estimating its original breadth, the line of its true centre should be marked on the copy. Such an observation will be invaluable to the restorer of the text. All letters that have peculiar forms should be carefully drawn at least once, as specimens. If the text is grouped about, among, or in any relation to, sculptures, such relation should be indicated, even if the whole sculpture be not drawn. Marks of punctuation or division, ligatures of one letter with another, and ornate initials and finials, should be looked out for and noted. On beginning a copy it is well to transcribe first the most obviously easy parts of the inscription, irrespective of order: they will teach the particular forms of lettering used, and accustom the eye to the inscribed surface before it has to do harder work. On finishing a copy, if there be time, read it over, try to translate it, and in the light of the probable translation, attack again the harder parts of the inscription. In the case of stones with two or more texts inscribed one over the other (as is frequently the case with Roman milestones, which may be geographical documents of great importance), a 'squeeze' must be got at all costs, for an untrained copyist will make little or nothing of them, and find it very difficult to draw an accurate hand fac-simile.

3. *Literary Memoranda.*

It is impossible to make too many notes of a monument, and quite easy to make too few. The nature of the great majority of such notes must be left to the discretion of the traveller; but concerning all antiquities, from buildings to beads, it may be said that, at any rate, material, colour, dimensions, condition of preservation, arrangement of parts and character of ornament must be jotted down in the ever ready note-book. In the case of buildings, notes giving all dimensions are especially important, since the camera can seldom be brought to bear on all parts and details, and there is often not opportunity to draw out a plan on the spot. Of inscriptions certain facts must be recorded, viz., form of the stone; whether its text be sepulchral, religious, civic, or what; the material and colour; on what sides complete, and on what not; actual dimensions; height of lettering; general character of lettering; whether well or ill cut, plain or ornate. Epigraphists' note-books are procurable in Germany and Austria containing divisions for the notes on all these points, together with a ruled space for the actual inscription. In the case of coins, the material, the ancient value, the weight (if possible), the state of preservation, and the images and superscriptions should be noted; of sculpture, the material, the dimensions, the degree of finish, and a minute analysis of the subject, the dress of the figure or figures, their gestures, attributes, and so forth.

B.—Cleaning and Conservation.

The traveller will not have occasion to render more than "First Aid" to objects of antiquity, *i.e.* to clean them so that their true character may appear, and to consolidate them with a view to safe transport. He should do the least that is absolutely necessary, leaving all elaborate treatment to experts at home. The hints here given are therefore elementary, and concerned mainly with portable objects. In this connection these are best classified by the materials of which they are made.

(*a*) *Gold.*—Gold requires no immediate treatment, unless it be laid over a core of other metal, *e.g.* copper or bronze, which has oxidised out through cracks. This oxide can usually be removed by picking or by sharp scaling blows with a small hammer and chisel. If it is obstinate, use dilute hydrochloric acid laid on with a brush. The proportion of pure acid should not be more than one part to ten of water.

(*b*) *Silver.*—If in very much corroded condition, technically known as 'dead,' silver should be left alone. If there is only slight superficial corrosion, soak in a solution of common salt or lemon juice, or strong ammonia, and polish after a few hours. If there is a good deal of chloride on the surface, but the body of the metal seems sound, put zinc or iron in the solvent and the chlorine will pass over, leaving a powdery surface which can be brushed clean. Silver should never be packed in a tin box, even if wrapped up, or it will be found on arrival at home to be stained with a brown rust, very difficult to remove. Pack in wood or cardboard.

(*c*) *Copper and Bronze.*—Both copper and bronze objects are best cleaned, if possible, with hammer and burin. Bronze, however, which the traveller will meet with most often, is frequently covered with a corrosion which cannot be scaled off at once, and calls for an acid solvent. Dilute hydrochloric (1 : 10) will act most quickly and effectually, but it leaves a white oxy-chloride coat, not easily got rid of either from the metal or the fingers of the operator, which it stains deeply. On the whole, we recommend that (1) if the bronze be badly cracked it be left alone; (2) if not cracked, but covered with very hard corrosion, soaking in a weak solvent like lemon-juice be tried; and (3) after that (or before, if the corrosion will yield to a tool at all) every effort be made to pick or flake off the corrosion, after which the object should be rubbed well with oil in the palms of the hands.

(*d*) *Lead and Iron* are both best left alone by the traveller.

(*e*) *Stone.*—The traveller will seldom or never be under the necessity to treat stone surfaces, except in the case of inscribed marbles on which carbonate of lime has formed, or small objects attacked by salt. The former can be cleaned with strong acid; the latter must be soaked in water for long periods, and when drying laid with the most important surface downwards, so that evaporation into the air may take place through the less important surfaces.

(*f*) *Pottery and Terra-cotta.*—The same enemies, carbonate of lime and salt, attack pottery and terra-cotta, and are met in the same way as in the case of stone; but the hydrochloric or other acid solvent should be weaker, and where there is colour be very cautiously used, if at all. In packing vases, it should be remembered that if many are put into one case without partitions, and one collapses, this will probably entail the breaking of the lot. Large vases should be filled with tightly rammed packing. Nothing of heavier or more solid material, *e.g.* stone, should be put in a case with pottery.

(*g*) *Wood, Ivory, and Bone.*—These materials are generally found flaked, split, or scaly, and need consolidation before packing. The methods most likely to be open to the traveller are: (1) if the objects are not excessively tender, to dip them in melted vaseline, let this set, and then pack away carefully in cotton wool; (2) if the objects are very tender or rotten with salt, to make a stiff jelly, drop them in before it sets, and convey them home in aspic. On the way the jelly will absorb the salt. In addition to either process, it is often well to bind the object in every direction and part with fine thread to keep it from splitting, or prevent the sections, if already split, falling apart and splintering or warping in different directions. If there is crystalline carbonate of lime on the surface of bone or ivory, it had better not be touched except by experts.

(*h*) *Papyrus.*—Papyrus needs damping and flattening out as far as possible without breaking its fibre. Lay it between two damp towels, and after flattening, pack it between sheets of paper in close tin boxes, filling each box up tightly. But the traveller will be wise not to try too much. As soon as the edges of his roll or fragment cease to be brittle to the touch he had best leave them as they are without further unrolling, and pack very carefully away.

(*j*) *Glass, Glazed Objects, Pastes, Amber, Various Compositions.*—None of these had the traveller best try to treat. Glass must be left wholly alone if in a flaking state; other objects, if flaky or powdery, will, at least, be no worse than they would be in any case, after transport, for having been dipped in melted vaseline.

No general hints can be usefully given as to either the methods of discovering antiquities or those of detecting forgeries. It is not supposed that the travellers here addressed will undertake regular excavations. Should they propose to do so, they will need special training and much more elaborate instruction. Mr. Petrie's 'Methods and Aims' will supply much of the latter, but the tyro excavator will not have Mr. Petrie's success without serving a long apprenticeship.

The above hints will serve for the traveller who takes antiquities by the way and in the day's work, having other objects more especially in view. For him the following list of necessities will suffice:—

1. Long extension camera and all photographic requisites, including, if possible, a stand and frame for overhead photography (*v.* p. 52).
2. "Squeeze" paper, brushes with hand-straps or curved handles, and tin cylinders for carrying the paper and moulds; heel-ball, sealing-wax, tin-foil, beeswax.
3. Planning and drawing outfit.
4. Notebooks ruled *en quadrille*, magnifying-glass and hand-mirror.
5. Small hammer and chisels for metal work, hydrochloric acid, ammonia, emery-paper, burin, plate brushes, sheet zinc, vaseline, cotton-wool (*not* raw cotton-waste with hard seeds in it).

VIII.

MEDICAL HINTS.

By the late WILLIAM HENRY CROSSE, M.D.

Revised by CHARLES FORBES HARFORD, M.A., M.D.

THE following hints, which were compiled by the late Dr. W. H. Crosse, formerly the experienced medical adviser of the Royal Niger Company, based upon previous editions of this work, have been edited so as to bring them up to date. The greater part remains as Dr. Crosse wrote it, but the section on Malaria has been almost entirely re-written, in order to bring it into line with modern views. The present editor naturally does not hold himself responsible for all the views expressed, but he believes that the advice given in these pages is sound, and will be found of inestimable value to travellers.

INTRODUCTION.

IN compiling these hints the author has followed as closely as convenient those written by the late Surgeon-Major Heazle Parke in the seventh edition of this manual, but to a large extent they have been re-written, and a considerable amount of fresh matter has been added. The space at the author's disposal being limited, the hints are but brief, and many things of importance have had to be omitted.

In the following pages the chief emphasis has been laid upon the care of the health in the tropics; but it must be remembered, that whilst by far the greater proportion of travellers go to the tropics, most of these hints for the preservation of health apply equally well for all climates.

Though many subjects have been briefly dealt with, certain matters have been rather more fully written up, such as the treatment of wounds. It is, in the author's opinion, so important that the traveller should thoroughly understand what is meant by "surgical cleanliness" that the usual methods observed by surgeons to ensure it have been plainly set forth. The author, of course, understands that in many cases it would be quite impossible to carry out the instructions in every detail, but it is hoped that a thorough knowledge of the principles underlying the correct treatment of wounds will assist the traveller in doing the best possible for his patient in any emergency, and under even the most unfavourable circumstances.

It is hardly necessary to observe that travellers in remote regions, and especially in tropical climates, are much more exposed to physical ills and diseases than most residents at home, and that they are more likely to be placed beyond the reach of skilled medical and surgical aid when it is most required. It is chiefly for the use of the non-professional traveller that the following pages have been written, and with this aim in view the symptoms and general treatment of the diseases and injuries with which he is most likely to be brought in contact are dealt with in simple, non-technical language.

Every traveller should supply himself with either Gardner's 'Household Medicine and Sick-Room Guide' (Smith, Elder & Co.), or with 'The Shipmaster's Medical and Surgical Help' (Griffin & Co.); or if a smaller book is desired, Russell's 'Domestic Medicine and Hygiene' (Everett & Co.), which appears to be one of the most suitable of all medical guides.

Persons who intend to travel should undergo a thorough medical examination, in order to ascertain if they are likely to be able to stand the fatigues, exposures and privations to which they will probably be subjected.

Most people of good constitutions and regular temperate habits can, with care, maintain a fair state of health in the tropics, and many of those who have been by no means strong at home are able with some extra caution to do well even in the hottest climates.

The intending traveller should attend a course of ambulance lectures in order to prepare himself for the responsibilities which he will have to undertake, and special lectures on the preservation of health in tropical climates are delivered from time to time by the present editor of these hints, who is the instructor in health and outfit to the Royal Geographical Society. He may be addressed at the office of the Society. As much time as possible should be devoted to discussing with some

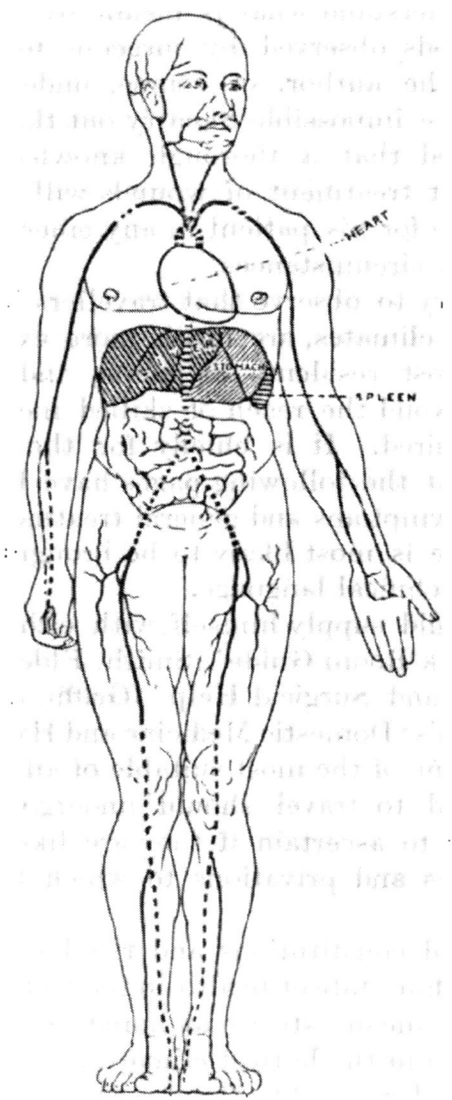

FIG. 1.—DIAGRAM SHOWING SOME OF THE PRINCIPAL ORGANS OF THE BODY, AND THE COURSE OF THE MAIN BLOOD-VESSELS.

professional friend the uses, correct doses, and proper methods of prescribing and combining the drugs which are to be taken on the journey, and in becoming familiar with the position of the principal bones, vessels and other important structures of the human body.

A traveller should understand how to use the clinical thermometer, how to twist or tie a bleeding vessel, how to use the hypodermic syringe and the syphon stomach tube. He should learn how to cleanse a wound, and should know the best way in which to set a broken limb. Such things are not easily learnt from books, and experience alone will give the necessary skill and confidence. Many valuable lives have been saved by travellers who have fairly mastered the rudiments of medical and surgical treatment, and who have been able in an emergency to give the necessary drugs, administer an antidote, or to stop severe bleeding.

The ideal traveller is a temperate man, with a sound constitution, a digestion like an ostrich, a good temper, and no race prejudices. He is capable of looking after details, *e.g.*, seeing that drinking water has *really* been boiled, and is willing to take advice from those who have made journeys in countries similar to the one in which he is about to travel.

A traveller should be supplied with suitable housing, food and clothing, and should have a proper supply of medicines, dressings, and medical comforts.

In recommending the drugs with which the traveller should be provided before leaving home, I have chosen chiefly the tabloid preparations of the firm of Messrs. Burroughs, Wellcome & Co., as after a considerable experience of tropical travel and exposure, which form the severest test of the reliability of medicines, I have found, like Parke, that they are the best in the matter of constancy and unchangeability of strength, and they have the great advantage of being extremely portable.

Risks to Health in a Tropical Climate.

Seeing that these hints will be chiefly required by travellers in tropical or sub-tropical climates, it may be well to indicate some of the chief risks which are to be met with in warm climates, and the best methods of meeting them.

It will be noted that any detailed reference to questions of outfit, such as the supply of proper food or clothing, is omitted, as these matters have been fully dealt with in 'Hints on Outfit,' published by the Royal Geographical Society uniformly with this 'Hints to Travellers,' and it is understood that this work will be consulted by all who read these hints.

Chill.—The most prominent feature of a tropical climate is undoubtedly heat, and yet the greatest risk arises not so much from excessive high temperature as from a risk of chill, which is due mainly to the excessive perspiration which takes place. This must be guarded against by the use of proper clothing, the most important point being that the underclothing should be of absorbent material. The form of underclothing chiefly recommended is either the Aertex Cellular Clothing or one of the different forms of woollen material. This question is fully dealt with in 'Hints on Outfit.' It is particularly important to avoid sudden changes of temperature, and if exposed to the cooling sea breeze, special care must be taken as to proper clothing. Parke laid great stress on the importance of avoiding chills, draughts, or wettings. He remarked "In crossing Equatorial Africa the Relief Expedition found that every wetting meant an attack of fever." The sea breeze, which is so refreshing and cool, by checking the perspiration, frequently acts as the exciting cause of an attack of fever. There is special liability to chill and subsequent fever when ascending an eminence, as the exertion causes profuse perspiration, and the cool breeze encountered on arrival at the top is very likely to produce ill effects.

On completing a day's journey, the underclothing, at least, should be changed without delay, and the skin should be well dried by the free use of a rough towel.

Effects of the Sun.—It is difficult to overestimate the importance of the protection of the head from the direct rays of the sun. It is best, where possible, to avoid going out in the heat of the day, but where this is necessary the head should be protected by a suitable helmet, which should be light in weight and colour, and which should have a good protection for the back of the head. A large green leaf inside the helmet might be an additional protection, and a sun umbrella should also be used wherever practicable. A helmet should always be worn when going out during the day-time in the tropics.

Errors of Diet.—The lassitude which is often so much felt by Europeans resident in the tropics too frequently tempts them to the abuse of alcoholic stimulants and highly-spiced foods. The habit is a most pernicious one, for such indulgence is one of the most fruitful causes of the permanent ill-health so often wrongly attributed to the mere residence in a hot climate. There is no doubt that food should be taken with greater moderation in hot than in cold climates; heat-producing articles of diet, such as fat, should be taken in far smaller quantity, but an ample supply of vegetables is essential. The meals should never be heavy, especially during the heat of the day, and intervals of about four hours should always separate consecutive meals.

Cooking should always be conducted with great care in the tropics, for the stomach and liver are less able to bear any extra strain, such as would be induced by the attempt to digest imperfectly-cooked food. Parasites are often introduced into the system by insufficiently cooked food.

Natural milk must be boiled; condensed milk should be mixed with boiled water.

Excess in the use of alcoholic stimulants is one of the most fatal errors into which the tropical resident can fall, and its use as a beverage is totally unnecessary, tea, coffee and cocoa being the best beverages for ordinary use. A small supply of brandy, champagne, and port wine may be of use in certain cases of illness, but they should be regarded as belonging strictly to the medical equipment.

Avoid native drinks, as they will probably have been diluted with dirty water, or prepared in unclean vessels.

Great moderation in the use of alcohol is quite as necessary in arctic as in tropical climates.

Risks due to Drinking Water.—The use of water for drinking purposes must be attended with great care in all tropical climates. As the water of the lakes, streams and pools of these countries usually contains a large proportion of impurities, and the germs of many diseases, it should be strained and subsequently boiled before being used.

As, however, it is not always possible or convenient to incur the delay of boiling the drinking water and allowing it to cool, it is advisable that a *reliable* filter should be taken.

Most filters—charcoal or otherwise—are merely death-traps, as the accumulation of germs and injurious matter within the filtering substance soon renders the water more dangerous than if unfiltered. There are, however, a few filters which, with ordinary care in cleansing, are in themselves efficient safeguards. The most highly recommended of these are the Pasteur Chamberland filter, supplied by Messrs. Defries and Co., 146 Houndsditch, E.C., and the filters made by the Berkefeld Filter Company, 121 Oxford Street. The latter company supply the most convenient filters for the use of travellers, and the field-service filter used largely by the Army in South Africa is the best type for an expedition. In any case, spare filtering candles should be taken, and these should be cleaned and boiled at least every three days. It is recommended that the filter-makers should be consulted immediately the probable requirements of the expedition are known.

The drinking of very cold water, to which there is great temptation when one is exhausted by prolonged heat and copious perspiration, should be carefully avoided; thirst often induces tropical residents to have recourse to iced water, which is always dangerous. The drinking of copious draughts of water is also a habit to be deprecated; it certainly weakens the muscular energy, and as the water is rapidly lost by perspiration, the feeling of exhaustion is increased. Hot or cold weak tea, without milk or sugar, is one of the least injurious of all beverages.

Precautions on the Voyage.

The traveller should endeavour to land in a perfectly healthy condition, and to this end he should on the voyage out take plenty of exercise, drink little or no alcohol, be moderate as to diet, and avoid much meat and rich dishes. Neglect of these obvious rules frequently ends in the traveller arriving in a flabby, bilious condition, in which state he is predisposed to attacks of malaria, dysentery, and other diseases. It is better to drink only mineral waters, as filters on board ship are often defective.

Constipation is frequent at sea, and a seidlitz powder, a dose of fruit salt, or one or two cascara tabloids may be necessary. If constipation is severe, then one or two four-grain blue pills should be taken at bed-time, followed in the early morning by a seidlitz powder or some other saline aperient.

In order to avoid chill the traveller should be properly clothed during the evening or when there is a cool breeze, and should not stand in draughty doorways and passages on board.

What is true of the journey out is of even greater importance on the voyage home. There is a serious danger of illness owing to chills contracted on board ship when leaving a hot climate. Quinine should be systematically taken as recommended on p. 170.

General Hints.

Never take a cold bath in Africa unless ordered to do so by a doctor.—In the case of persons who have already suffered from many attacks of fever, dysentery, or any disease of the liver or other important organs, warm bathing should alone be used. Bathing should never be resorted to during the period of digestion, i.e., three to four hours after meals. Wear warm clothes at night. Avoid the direct rays of the sun. Do not take too much animal food. Never begin work on an empty stomach. Never neglect a slight attack of fever or diarrhœa. Keep the bowels gently opened—once a day is quite enough—but avoid strong purgatives.

Vaccination.

It is essential that all the members of an expedition should be re-vaccinated if they have not suffered from smallpox or been vaccinated within two years. A proper supply of vaccine should be carried on the journey. Dr. Chaumier's calf lymph, sold by Roberts and Co., seems to be the most reliable for use in hot climates.

Isolation.

On the outbreak of an infectious complaint, such as smallpox (which is very common amongst negro carriers), the sufferer should be promptly isolated, and one or more attendants should take charge of the case and not be allowed to come to camp. It is wise to select as attendants those who have either had the disease, or who are protected (as by vaccination) against it.

In Nigeria it was found most convenient to build grass huts, which were burnt to the ground, together with the clothing and bedding of the patient and attendants, on the termination of the outbreak.

All vessels which are not destroyed must be boiled thoroughly. The motions, etc., of patients suffering from typhoid fever, cholera and dysentery should be disinfected, or, better still, destroyed by fire.

Diseases and their Treatment.

Malaria.

The risks to health which may be attributed to the food supply, to drinking water, or to chill may be met with in almost any climate, though there is a special liability to diseases from these causes in the Tropics. There are, however, certain diseases prevalent in the Tropics which are conveyed in a manner quite different to that of the ordinary diseases met with in a temperate climate. Many of these diseases, including malaria, yellow fever, sleeping sickness and probably plague, are conveyed by the bites of insects, and in the case of the first two of these it appears to have been proved conclusively that protection from the bites of infected mosquitoes gives absolute freedom from the diseases in question.

When it is recognised that malaria has proved in the past the chief obstacle to the satisfactory development of tropical countries, it will be readily seen how important it is that the traveller who is proceeding to malarious districts should understand the risks which he must face, and should take steps to protect himself from these terrible diseases.

In order that this may be satisfactorily done it is most desirable that each traveller should understand at least the chief facts concerning the propagation of malaria by the mosquito, and good information on this point, as well as upon the prevention of other tropical diseases, may be obtained from a little book by Dr. W. J. Simpson, entitled "The Maintenance of Health in the Tropics," published by John Bale, Sons and Danielsson, Limited. It may be well, however, to summarise in a few words the chief facts concerning the propagation of malaria.

How Malaria is Propagated.

1. Malarial fever is caused by a small parasite which lives chiefly in the blood of patients attacked by it.

2. It can only be conveyed from one person to another by the bite of a mosquito which sucks the blood of the infected person.

3. The parasite undergoes a series of changes in the mosquito's body, which eventually finds its way to the poison glands of the mosquito and is then injected by the mosquito into its next victim, and it is only in this way that a person can be infected by malaria.

4. It is only a certain kind of mosquito, known as the Anopheles, and only the female of this species, which can carry malaria, and this mosquito breeds in shallow puddles and in almost all collections of stagnant water. It is therefore very dangerous to pitch camps near to stagnant pools or marshy places.

5. It has been found that the blood of native children, especially within the first few years of life, swarms with malarial parasites, and it appears to be largely from the mosquitoes which come from the huts in which these children live that the infection is derived. It is therefore very unwise to camp quite close to ill-kept native huts or to go into a native village at night unless protected from the bite of the mosquito.

Full particulars concerning the best methods of protection from mosquitoes will be found in 'Hints on Outfit,' Chapter V., but certain points may here be emphasised as to the practical steps which should be taken for the prevention of malaria. As these hints are intended only for travellers and explorers, and not for the ordinary resident in tropical climates, it is useless to describe at length the measures which may be adopted for the extermination of mosquitoes in the neighbourhoods of human dwellings, but as already advised it is essential that the camp should be so placed as not to be near to the breeding places of Anopheles. It is clear then that the chief aim of the traveller must be to protect himself from the bites of mosquitoes.

Methods of Prevention.

Protection from Mosquitoes:—

(*a*) It is specially important that those who are suffering from malaria should be protected by suitable mosquito curtains from the bites of mosquitoes, not only for the comfort of the invalid, but to prevent the disease from being communicated to his companions.

(*b*) Adequate mosquito curtains should be used at night in the way described in 'Hints on Outfit,' and it should be realised that a mosquito net is absolutely useless unless it is thoroughly efficient, and constant watchfulness is therefore necessary, not only to see that the mosquito net is of suitable pattern, but that it is efficiently used.

(*c*) Whilst it is undoubtedly important to protect one's self during the hours of sleep, it is perhaps in the evening time after sunset, and before going to bed, that most people are badly bitten. It is therefore highly desirable to have a mosquito canopy, in which it is possible to sit with comfort during the evening, and this will not only protect from mosquitoes, but from flying ants and beetles and many other winged insects, and may even be some protection against lizards, centipedes, and scorpions. It is even possible to arrange a mosquito canopy in which the evening meal can be taken. In this case one servant boy should be inside the mosquito canopy and receive the dishes through a form of mosquito trap from the attendant outside.

(*d*) As, however, it may be difficult to arrange to dine under a mosquito canopy, especially if there are several together, it is important to take special steps to protect the ankles during the evening. This can be done by the use of mosquito boots, which are described in 'Hints on Outfit.' The mosquito boot consists of an ordinary shoe or slipper, with a prolongation up the leg of soft leather, after the pattern of a long boot, though without the pressure on the leg which would be caused by a stiff boot. They would not, of course, be needed when within the mosquito canopy. If cane-seated chairs are used, or with perforated seats, a cushion should be placed on the seats. Those who have to travel at any time by night, or who have to go about after dark, perhaps for purposes of sport or for other reasons, should protect their face by mosquito netting attached to the helmet and tied round the face, and should wear gauntlets for the hands.

The Use of Quinine.—Whilst it should be the endeavour of the traveller to prevent inoculation with the malarial parasite, it is difficult to absolutely avoid infection. It is therefore most desirable that each individual should take quinine regularly, as quinine acts as a direct antidote to the parasite in the blood, and usually prevents the development of the symptoms of malaria, even though the parasite has been introduced into the system. Perhaps the simplest plan is to take 20 grains of quinine a week, which means that 5 grains should be taken on Monday, Wednesday, and Friday mornings, and another dose on Sunday. It is particularly necessary for this to be kept up after an attack of fever has occurred, and those who have had malaria in the tropics should continue the taking of quinine for two or three months after they return home, as serious attacks of malaria often take place during this period. The quinine may be taken in the form of tabloids of either the bi-sulphate, hydro-chlorate, or hydro-bromide of quinine, or, if these forms of quinine seem to be unsuitable, Euquinine may be tried, though it is more expensive than other preparations. The action of quinine, when dissolved in water, is more rapid and certain than when taken in pill or tabloid form, but the convenience of the latter is so great that it is probably most generally satisfactory.

Symptoms.—Malarial fever presents itself under two forms. (1) *Intermittent* fever. In this disease the temperature may rise high, but returns each day to normal or lower; hence there is, after each attack, a period of complete freedom from fever. An intermittent fever or ague is usually less serious than a remittent fever. (2) *Remittent* fever. In this the temperature, though it varies, keeps constantly above the normal, and the higher the fever, and the slighter the difference between the extremes of temperature, the more serious is the condition of the patient. Where the temperature is remittent, and appears to be unaffected by quinine, the disease is probably not malarial, but may be a case of enteric fever, and should be treated as advised below under that heading.

The attack may be sudden, but it is usually preceded by a feeling of languor, yawning, and general discomfort; this is followed by the *cold stage*, which, in the tropics, is usually short, and in the more ordinary attacks is ushered in by a violent shivering fit or rigor, though this is not common in Central Africa. Then comes the *hot stage*, often of long duration, followed by the *sweating stage*. After this there is a period of remission, or intermission of the feverish symptoms; usually, after some hours, the attack comes on again, beginning with the cold stage, but if the fever is treated very early, the disease may now pass off.

As a rule, when a patient is suffering from malarial fever, the attack returns every day, but in some cases it only returns every second or third day.

Treatment.—The three great principles of treatment are: (1) To open the bowels; (2) to produce perspiration; (3) to give quinine. The routine to be adopted is as follows. Put the patient to bed in flannel pyjamas, and covered up well with blankets; if the bowels are not freely open give an aperient, such as four grains of calomel, or two tabloids of Cathartic Co, or two Livingstone's Rousers, and if this does not act, repeat the dose in about four hours. If free evacuation is not produced, a warm-water enema should be given. A hot-water bottle in bed is useful. Sponging with warm water often gives relief at the beginning of a fever, and tends to shorten the cold stage. At the same time hot drinks should be given, such as weak tea, in order to promote perspiration and cut short the hot stage, and at this stage, *i.e.*, at the outset of the fever, 15 grains of Antipyrin, or 10 grains of Phenacetin, may be found useful for inducing perspiration, or four tabloids of Warburg's

Tincture may be substituted for them. Antipyrin and Phenacetin should not be given except in the earliest stages of the fever.

Whilst the above methods of treatment are important, it should be clearly recognised that the one drug which can alone counteract the malarial affection is quinine, and it is upon the proper administration of quinine that successful treatment largely depends.

The best time to give quinine is at the commencement of the sweating stage, when a dose of ten grains should be given either in the form of tabloids, or if fever is severe, in solution. If, however, the sweating stage is deferred longer than four hours after the commencement of the attack, a dose of quinine should still be given, and repeated every four hours during the first day; but it should not be given during the night, and not more than thirty grains should be given during twenty-four hours. Thirty grains of quinine should be given for the next two days in doses of ten grains, but if it is clear that the fever is periodical, occurring at the same time every one or two days, a dose of fifteen grains should be given about four hours before the fever is expected, as this is the best way to prevent a paroxysm. This treatment should be kept up as long as the fever lasts, if it appears to be clearly a case of malaria; but if, as is most probable, the fever subsides at the end of two or three days, ten grains of quinine should be taken each day for a week, and after that the regular prophylactic dose should be reverted to.

Patients frequently on the second or third day leave off quinine altogether, and then have a relapse which they mistake for a fresh attack, and so they go on for weeks or months, constantly having "slight goes" of fever, till eventually, when anæmic and broken down, they suffer from a violent attack of malaria, perhaps associated with blackwater symptoms.

If during the attack of fever the quinine cannot be retained by the stomach, it must be administered by the bowel, remembering that twice as much must be given by enema as would be given by the mouth. When vomiting is persistent, it is best to make the patient drink warm water freely, half a pint to a pint or more, which will be vomited, and may thus remove irritating matter from the stomach; then give him a sedative draught containing twenty drops of chlorodyne. Five or ten grains of bismuth subnitrate may assist in allaying the vomiting. A large warm-water and soap enema will frequently check vomiting by increasing the action of the bowels, and a mustard leaf may be applied to the pit of the stomach. When other remedies fail, small quantities of champagne may help to check vomiting.

Drinks of diluted lime-juice and water will be found refreshing, but if the lime-juice is given too strong it may cause vomiting.

Cloths kept wet with an evaporating lotion made of one ounce of spirit in half a pint of water may be applied to the head, and cold sponging of the face, neck, arms and chest helps to lower the temperature, is very refreshing, and can do no harm if the patient is not allowed to get chilled; if he is very feverish, it may be necessary to give a cold bath. See further details below.

During the sweating stage care must be taken that the perspiration is not stopped by a chill. Hot drinks will help to make the skin act freely.

The temperature is the best guide to the treatment, therefore it should be recorded and kept for comparison with subsequent temperatures.

If the patient is very uncomfortable during the attack of fever, twenty drops of chlorodyne may be given with advantage.

If there is much pain in the left side under the lower ribs, *i.e.*, in the position of the spleen, hot fomentations or poultices may be applied.

If the urine is irritating, or the quantity passed is reduced, give plenty of fluid, and administer five grains of bicarbonate of soda three times a day.

As soon as the patient can stand it, that is, when the fever is nearly gone, tonics, such as iron and arsenic, will be required.

Diet.—The patient must be fed between the attacks of fever, or when the sweating stage is on, with good soup, slightly thickened with sago or rice, or with eggs beaten up with milk. If it is difficult to administer nourishment and the pulse is very weak, one teaspoonful of brandy may be given to two tablespoonfuls of warm milk.

If the attack of fever is prolonged careful feeding is essential, as advised below under the head of Blackwater Fever.

If the patient cannot take food, then a nutrient enema must be given.

Baths, etc.—In ordinary cases it is not necessary to take special measures to reduce a temperature of 102° or 103° F. unless it is kept up; but if the temperature reaches 104° or 105° F., and remains there, cold applications to the whole of the body will be required. (*See* Baths, p. 258.)

In cases where there is acute malarial poisoning, with temperature rushing up to 105° or 106° F., or even higher, do not wait to undress the patient or get a bath, but empty gallons of water over him, one boy keeping the head constantly soused; while this is being done a bath can be procured and the patient then undressed, or, better still, have his clothes cut off, because it is dangerous to lift such a patient about too much. Remember that when a patient is very ill and weak, he should not be allowed to stand or sit up suddenly, as he may faint.

Delirium with high temperature, say 104° to 105° F., is a certain sign that the fever is doing harm, and must be reduced.

Do not trust too much in drugs; cold water is absolutely necessary.

Death from simple malarial fever is hardly known; patients should be told this.

Blackwater Fever.

Blackwater fever is probably a pernicious form of malarial fever, and derives its name from the colour of the urine. It must be remembered that dark-coloured urine is usual in all fevers; it is scanty during the height of the fever, especially if there is much sweating. If, however, it is obviously "bloody," the case is more grave, but as a rule it is only men broken down in health, and those who have resided in tropical Africa for at least a year, who suffer from this complication.

The reason for the occurrence of this condition is not known. Some have ascribed the symptoms entirely to the taking of quinine, but as the fever often occurs where no quinine has been given this is impossible. At the same time it appears likely that in certain cases of malaria, owing perhaps to some idiosyncrasy, quinine may help to bring on the symptoms. It is possible that chill occurring during the course of a fever may lead to the production of blackwater fever. Those who have had one attack are particularly liable to a recurrence, and after two consecutive attacks return to a temperate climate is required.

Symptoms.—In addition to the ordinary symptoms of malarial fever, the urine is dark, blood-like, and eventually porter-coloured; it is often scanty, and may become entirely suppressed. The skin is yellow, often a bright orange, there is frequent vomiting, and the vomited matter is usually of a green colour.

Exactly the same treatment should be adopted as that fully described above for malaria. It is especially important to give an aperient at the beginning, and perhaps five grains of calomel is the best form.

The chief aim should be to support the strength by fluid nourishment, and to secure free action of the kidneys. The former should be maintained by fluid nourishment given in small quantities at frequent intervals, such as milk, Plasmon, Benger's food, Allenbury's foods, invalid Bovril, Brand's fever food, or Brand's essence, Maggi's consommé. A little Plasmon added to any of the meat preparations would be useful.

For the rest, reference must be made to the treatment of malaria. The indications for the administration of alcohol would be the same as mentioned under that heading.

In order to maintain free action of the kidneys, plenty of fluid should be given, such as barley-water made from Robinson's prepared barley, flavoured slightly with lime-juice or lemons.

Vomiting is often a serious complication, and the directions for its treatment, given under the head of malaria, should be carefully followed. If it cannot be speedily checked, feeding by the bowel must be carried out. *See* Nutrient Enema, p. 241. In between the administration of the nutrient enemata, the injection of one pint of salt solution (one teaspoonful of common salt to the pint of hot water) should be slowly and carefully injected into the bowel as often as it can be tolerated.

So long as plenty of urine is passed and sufficient nourishment is taken there is little cause for anxiety, though wherever possible skilled assistance should be obtained at the earliest opportunity.

Tick Fever.

One of the diseases which may be produced by the bites of insects is a form of fever, conveyed by the bite of a small tick, which is common in many parts of Africa. This produces a series of symptoms which are similar to those found in the disease known as relapsing fever, which has been known to occur even in the United Kingdom, and which frequently occurs in cases of famine.

Symptoms.—The chief symptoms are those of a severe attack of fever, ushered in by a shivering fit and acute symptoms such as are usually found in cases of fever, such as pains in the back and limbs, rapid pulse, and sometimes severe vomiting and diarrhœa. The fever generally keeps up for about a week, after which there may be an interval without fever for a few days, to be followed later on by another attack of fever.

Treatment.—The important point is to guard against being bitten by ticks, which can best be done by the use of a ground sheet and mosquito net when camping out or in native dwellings, if it is necessary to sleep in them. The disease is very likely to be mistaken for malaria, and except that quinine is not likely to affect the course of the disease, the treatment given may be such as is used in the more severe forms of malaria. It is most important to support the strength of a patient during the continuance of the fever, and quinine may be given, though it is not likely to very greatly modify the progress of the disease.

Sleeping Sickness.

This disease has lately proved to be one of the most terrible scourges which are to be met with in Central Africa. Until recently it was only known on the Congo, whilst isolated cases occurred throughout the west coast of Africa. It has recently however spread to Uganda, where it has almost depopulated certain regions. It is due to the bite of a tsetse fly, somewhat similar to that which produces the terrible disease of horses, which introduces into the system a minute blood parasite.

Symptoms.—Lassitude and sleepiness, associated with slowness of movements and speech, are among the early symptoms, which gradually increase until the patient sleeps continuously for hours, only waking up for meals; finally the disease terminates by coma, often convulsions, and death. It appears to be always fatal.

Treatment.—Various kinds of treatment are being tried, but none has been definitely proved to be efficacious. The most important thing is to avoid the bite of the tsetse fly.

Diarrhœa.

Diarrhœa, or looseness of the bowels, is one of the most common and one of the most serious ailments of the tropics, and should never be neglected. In many cases it is a sign of enteric fever, dysentery, cholera, or sprue, the symptoms of which are given below. Ordinary attacks are usually due to the presence of some irritant in the bowels, such as bad food, unripe fruit, or other poisonous material.

Treatment.—Begin the treatment by administering castor oil, fruit salt, cascara, or other mild aperient, to clear out the cause of the diarrhœa. A warm-water enema of about a pint is useful.

If the diarrhœa continues, give chlorodyne (20 minims) and tincture of ginger (10 minims) in an ounce of water two or three times a day.

This treatment should not be persisted in for more than two days.

If the diarrhœa is persistent, an astringent is needed: five grains of tannin, or two or three grains of sulphate of iron may be given three times a day. Ten grains of quinine should be given each day.

All food should be semi-solid and tepid; milk diet, as recommended below for enteric fever, is the safest diet, but soup thickened with rice or arrowroot is good. Patient should keep in bed and wear a flannel band round the belly.

If the trouble continues for more than a few days it is probably due to dysentery, or typhoid.

Enteric or Typhoid Fever.

It is impossible to give a full account of enteric fever here, and moreover the presence of a medical man is absolutely necessary for its proper treatment. The chief early symptoms, however, will be given, and a few hints as to their treatment.

The disease is characterised by ulceration of the small bowel, with continued high fever, and is usually accompanied by diarrhœa.

Causes.—It is generally caused by the drinking of impure water, but may also be contracted from bad drains, infected clothing or milk.

The incubation period is from ten to fifteen days.

Symptoms.—The early symptoms of the disease are often so slight that the patient will not believe he is really ill; he may just feel out of sorts, or complain of headache, but still go about his work. There may be diarrhœa, or occasionally constipation (*see* note on Constipation, p. 183). After five or six days the patient is generally compelled to give up and go to bed, headache or diarrhœa, or both, being the chief complaint. The temperature goes up in a characteristic manner, rising a little more every evening till it eventually reaches 103° or 104° F. There may be some cough, and often this symptom is a very troublesome one.

The belly is usually distended and slightly tender, and there may or may not be the characteristic typhoid rash, consisting of rose-pink circular, slightly raised spots, about the size of a large pin's head. They occur chiefly on the chest and abdomen, and come out on successive days, often only three or four at a time. These spots are frequently absent, and then one must be guided by the presence of other symptoms.

The possibility of enteric fever should always be remembered in cases where there is constant fever, unaccompanied by any definite symptoms, such as the recurrent shiverings of malaria, or the spitting of blood in pneumonia.

Treatment.—Absolute quiet in bed. If constipated, bowels should be kept open by soap and water enema only.

Milk only (three to four pints daily) should be given during the whole course of the illness and till ten days after the temperature has descended to, and remained, normal. Stimulants, if pulse is feeble and rapid; opium if there is much pain. If bleeding occurs from the bowel, an ice poultice or cold-water cloths should be applied to the belly; ice may be given to suck; opium and an astringent, such as tannin, administered by the mouth, or an opium enema be given. Milk should be given in small quantities only, and to each half-pint five grains of bicarbonate of soda should be added.

The motions must be burned, or carefully disinfected.

Dysentery.

Dysentery, or inflammation of the lower bowel, is caused by drinking impure water (dirty filters are a fruitful cause of this disease), and by eating bad or improperly cooked food; it may be provoked by chills, general debility, and exhausting diseases, such as chronic malaria. Alcoholic and other excesses render people peculiarly liable to contract the disease. Dysentery may complicate or be complicated by malaria.

Symptoms.—Diarrhœa with pains in the belly, straining and frequent desire to go to stool. The motions soon become small in amount, slimy, lose their natural colour, and contain more or less blood; when there is ulceration of the coats of the bowel, the motions are extremely offensive, and bleeding may be very free. There is heat, tenderness, and bearing down about the outlet of the bowel, with considerable prostration and probably some fever; there is frequently a constant desire to pass water. All these symptoms may be due to severe ordinary diarrhœa; but in the tropics it is best to treat them as if they were dysenteric.

Treatment of dysentery:

Essentials: rest, warmth, and suitable food.

Put patient to bed, apply a cholera-belt, and give a gentle purgative if the case is seen early. Take the temperature; if it is not high, give five grains of quinine twice a day; if high, give ten grains twice a day.

Diet.—Give as little food as possible during the early stages of the disease. Milk, or, better still, milk diluted with an equal quantity of barley- or rice-water, and just warmed, may be administered one table-spoonful at a time. Beaten-up eggs, and soups thickened with arrow-root, sago, or tapioca, may also be used.

The further treatment of the case consists in the administration of ipecacuanha, magnesium sulphate, or calomel. Ipecacuanha or sulphate of magnesia should be tried first, but if after a week the patient is no better, treatment by calomel must be adopted.

The object of treatment is not to block up the bowel—as might be done by giving large doses of opium or tannin—it is to cure the disease of which the looseness is only one symptom. The most favourable sign during an attack is a return of the colouring matter to the motions; this shows that the liver is again acting, and that the treatment is doing good. With the return of colour (which at first may be intermittent), the other symptoms, such as pain and bloody discharge, will abate, and the motions will become more solid and healthy.

In dysentery, as in severe diarrhœa, the patient should *not* be allowed to get up to stool. A box cut across obliquely will make a rough slipper bed-pan; put sand in it, and pad the edges. The dysenteric motions must be burned or thoroughly disinfected.

Treatment by Ipecacuanha.—When the bowels have been opened, give twenty grains of ipecacuanha, either solid or mixed with a wineglass of water, or less; arrowroot, starch, or gum-water, which will help to suspend the drug. Of course, ipecacuanha will act more quickly if it can be taken suspended in a liquid, instead of in the solid form. To prevent vomiting, put a mustard leaf to the pit of the stomach. Absolute quiet must now be observed; darken the room, and allow no moving in bed or

N 2

talking. Withhold food and liquid for at least two hours if possible, but if there is much thirst, teaspoonful doses of water may be given.

If there is no vomiting for an hour, probably a good part of the ipecacuanha has been digested; if it has been vomited, wait for half-an-hour, and then give another full dose. If vomited again, wait for two hours, and give twenty drops of chlorodyne, followed by twenty grains of ipecacuanha; the chlorodyne is to quiet the stomach, enabling it to retain the ipecacuanha. In about twelve hours from the first dose, repeat it in exactly the same way. If thirty grains are too much at a time, give twenty, three times a day, for not less than sixty grains should be given in twenty-four hours. The drug is not a dangerous one, and, if the patient can take it, too much can hardly be given. Between the doses feed the patient, giving but little at a time. If the ipecacuanha is going to do good, marked improvement should be apparent in four or five days; failure of the drug is often due to its not being given or retained in sufficiently large quantities.

If ipecacuanha cannot be retained by the stomach, it must be given by the bowel. (*See* Ipecacuanha Enema, p. 242).

Let the patient lie quiet so as to retain the enema. Up to four enemata may be given in the day. Too much ipecacuanha cannot be given, but the effect of the opium should be watched, and if more ipecacuanha is wanted, administer it in the form of tabloids which do not contain opium.

Sedatives, such as laudanum or chlorodyne, are of use only so far as they relieve pain, sickness, and great distress; the *full* dose in an ordinary case is twenty drops three times a day, but if less is sufficient, so much the better. Drowsiness is a sign that the patient has had as much as is good for him; no patient should take more than twenty drops at a time, unless it is ordered by a doctor.

Poultices, mustard plasters, and hot fomentations to the belly, do good by lessening the congestion of the bowels and liver.

If the lower bowel is uneasy, then a small enema, say of ten ounces of warm water, may afford considerable relief, or a *soothing* enema may be administered. (*See* Enemas, p. 242).

As the disease begins to abate, reduce the ipecacuanha and give tannin, five grains or more, three times a day; then half that amount. If diarrhœa is very profuse, bismuth or tannin may be given during the attack to assist the bowel by their astringent action. Sulphate of iron, or the solution of perchloride of iron, are useful when the acute symptoms subside, as they are astringent and tonic.

Bicarbonate of soda, five grains or more, dissolved in an ounce of water, is useful, as it allays irritation of the stomach; it may be given two or three times a day from the first.

Treatment by Epsom Salts (Sulphate of Magnesia).—This is an alternative treatment to that by ipecacuanha. The guide to the amount of salts necessary is the frequency of the motions caused by them, and the amount of relief of the painful symptoms. After one or two motions caused by the salts, it is noticed that the patient has not the same constant desire to go to stool, and that fewer motions are passed; that they are more copious, and gradually become a better colour. The object is not to violently purge the patient, but to keep the bowels gently acting.

Recently, a great number of cases of dysentery have been treated with marked success by giving sixty grains of Epsom salts every four hours; sulphate of soda may be used instead of Epsom salts and is less irritating. This treatment cures by causing constant free action of the bowels, and the consequent immediate removal of poisonous material.

When the motions become yellow, then an astringent such as iron is indicated.

Mercurial Treatment.—When the treatment by ipecacuanha or Epsom salts fails, then the administration of calomel may effect a cure.

One grain of calomel should be given three times a day for four days; if no unpleasant symptoms arise, and if the patient is improving, this dose may be continued for a week or ten days.

The addition of one grain each of ipecacuanha and opium to the calomel has been suggested by Dr. Manson. The objection to the calomel treatment is that mercurial poisoning may be produced if too large doses of the drug are given, or if its use is continued for too long a period; but with the doses mentioned here, this possibility is very unlikely.

At the same time the patient must be watched for such symptoms as tenderness of the gums, salivation (*i.e.*, great increase of the saliva), and a metallic taste in the mouth. Should these symptoms arise the dose must be reduced, or the drug discontinued.

Chronic Dysentery.

The treatment is to remove the patient to a healthy climate, and, meanwhile, to keep him as far as possible on a diet similar to that advised for acute dysentery. Opium and bismuth may be given to relieve pain, and tannin or sulphate of iron to check diarrhœa. Ipecacuanha, in ten-grain doses, three times a day, should be tried. An enema containing a teaspoonful of alum, or ten grains of sulphate of iron, to the pint, often does good.

An ipecacuanha enema often does more good than ipecacuanha given by the mouth.

Cholera.

Cholera is a serious acute disease, characterised by frequent watery motions, vomiting, cramp and collapse.

Cause.—It is usually contracted by drinking contaminated water.

Symptoms.—Giddiness, faintness, persistent vomiting and diarrhœa, great prostration, feeble pulse, cold perspiration, colic, intense thirst, and constant desire to pass urine. The vomit and motions rapidly become like rice-water in appearance, and the urine is more or less suppressed. There are severe cramps in the legs, belly, and other parts of the body. If then the pulse becomes weak, the temperature low, and the countenance dusky, the patient will probably sink. On the other hand, reaction may set in, all the symptoms abating, and the pulse, temperature, and colour becoming natural; the water is passed more freely, vomiting is less frequent, and the motions become more natural in colour.

Treatment.—Isolate the patient, keep him warm, and give ice to suck. Apply hot bottle to the feet, and mustard leaves to the pit of the stomach.

Give one drop of carbolic acid, together with twenty drops of spirit of camphor (or peppermint, or a little brandy), five grains of bismuth, and ten grains of soda, suspended in one ounce of gum water, every four hours. Chlorodyne may be given to allay severe pain.

In the early stages, food, such as milk, must only be given in very small quantities. All water used for drinking purposes must be boiled.

The motions, etc., must be thoroughly disinfected.

Sprue.

This is a chronic form of tropical diarrhœa, the leading symptom of which is the frequent passage of large, frothy, and pale-coloured motions; dyspepsia, and soreness of the tongue and mouth, are also present, and there is marked anæmia and advancing debility. The disease may follow attacks of diarrhœa or dysentery.

Treatment.—Put the patient to bed, give a simple aperient such as castor-oil, to clear out the bowels; allow only a milk diet, to which, as the symptoms begin to abate, meat juices and jellies should be cautiously added. If any other disease, such as scurvy, is present, it must be treated. Drugs are not usually of much service; however, a mixture containing bismuth, soda, and one drop of carbolic acid in an ounce of gum-water may be given three times a day for a week. The soreness of the mouth and tongue may be treated by the application of borax and glycerine, or mild antiseptic mouth washes, such as a weak solution of permanganate of potash, or a lotion of boric acid. Special symptoms, such as pain and collapse, must be treated as they arise. As soon as the patient is strong enough, he should be removed to a temperate climate.

Constipation.

This condition is very frequent in tropical climates, where it is associated with sluggishness of the liver. One of the best remedies is the two-grain tabloid of cascara, of which one may be taken three times a day. In addition to this, an occasional dose of a saline purge should be used, or a large enema of soap and water may be given. Five grains of blue pill or three grains of calomel will be found to act as a very efficient aperient, especially if followed in about six hours by a saline such as fruit salt. Castor oil in ounce doses is also a valuable remedy.

As a rule, the general health of people suffering from simple constipation is not seriously affected.

In cases of chronic constipation one of the anti-constipation tabloids, otherwise known as the Aloin Co., given three times a day and gradually reduced, will be found useful.

In most acute diseases, such as malaria, pneumonia, etc., if constipation is present, it should be treated at once by means of suitable aperients.

Note.—In peritonitis, *i.e.*, inflammation of the bowels, hernia, and in some cases of typhoid, constipation is a leading symptom, and is accompanied by severe pain in the belly. On no account should an aperient be given by the mouth in these cases. If it is necessary to clear the bowels, this should be done by means of an enema.

Colic.

This is the name given to the well-known severe twisting or griping pains in the belly, usually due to excessive flatulence, and resulting from constipation, or some error of diet.

Treatment.—Hot fomentations should be applied to the belly, or better still, the Instra, which is the best means of applying continuous heat to any part of the body, and a turpentine enema (a tablespoonful to a pint of warm water) will nearly always cut short the symptoms; in the absence of turpentine, give warm water alone. A full dose of opium (20 minims) should also be given if the pain is severe, preferably in a tablespoonful of castor oil.

Bicarbonate of soda, carbonate of ammonia and ginger should be freely given in full doses, and the bowels should be well opened as soon as the severe pain has passed off.

Congestion of the Liver.

The *Liver*, which is mainly on the right side, lies below the right lung, and is protected by the lower ribs. In health it extends vertically from one-and-a-half inches below the right nipple to the lower edge of the ribs; in certain diseases it is enlarged, and its edge can be felt well below the ribs.

Congestion of the liver is frequent in the tropics, and is often due to malaria or dysentery. Very frequently it is caused by abuse of alcohol, over-indulgence in food, and the excessive use of hot condiments, or by constipation and want of exercise. In the tropics the liver is more easily affected by excesses than in temperate climates.

Symptoms.—A furred tongue, sallowness of the face, headache, lassitude, disinclination for work, loss of appetite, tendency to vomit, occasional slight jaundice, and a sense of oppression about the region of the liver.

Treatment.—Light diet, abstinence from alcohol and spices, and the use of calomel or other aperient will usually effect a cure. Ammonium chloride, five to ten grains, three times a day, should be given.

Acute Inflammation of the Liver.

In this complaint there is severe pain, some fever, and frequently jaundice. The complaint is serious, as abscess of the liver frequently follows it.

Treatment.—The patient should be put to bed, hot fomentations applied to the seat of pain, and the bowels well relieved. Quinine will be required.

Ten grains of ipecacuanha should be given three times a day. Ammonium chloride in full doses (ten grains or more three times a day) often does good, and can be retained when ipecacuanha cannot, but it is unpleasant to take, so the dose should be given in one or two ounces of water. The wisest course for one who has had a severe inflammation of the liver is to get away to a healthy climate.

Abscess of the Liver.

It is extremely difficult for the traveller to decide if abscess of the liver is present; it may be suspected if a patient, convalescent from dysentery, still remains feeble and ill, or if he has an irregular temperature, night sweats, and wasting. Sometimes a definite swelling can be made out.

Treatment.—Ammonium chloride may be given; the patient's strength must be supported, and he should be as quickly as possible placed under the care of a surgeon, who will probably decide to operate. If this cannot be done, then the patient should not be interfered with surgically, for he will have a better chance of recovery if the abscess is allowed to burst naturally than he would if the traveller attempted to operate.

Yellow Fever.

This is a highly infectious disease, one attack of which usually protects against a second. It is conveyed from one patient to another by the bite of infected mosquitoes of the genus *Stegomiya*.

Incubation period, two to five days.

Symptoms.—The onset of the disease is very sudden, the highest temperature being reached almost at once; then follows a period of remission or calm, the pulse becomes abnormally slow, and this stage is usually either succeeded by convalescence, or the symptoms become worse and the patient dies. Some of the symptoms much resemble malaria, but the rapidity of the onset, severe pain in the forehead, eyes, and loins, the early scantiness of the urine, the bright eyes, the narrow red tongue, and the absence of pain about the spleen are characteristic.

There is considerable thirst and vomiting, and in bad cases the vomit becomes black, the colour being due to the presence of blood. (In ordinary malaria the vomit is yellow, or in severe cases, such as black-water fever, it may be of bright or dark-green colour.) In yellow fever, jaundice is developed about the third day, and tends to increase, whilst in blackwater fever it comes on very early and soon begins to abate.

Treatment.—Isolate the patient, and carefully protect from the bites of mosquitoes for fear of spreading the disease; open the bowels well by means of calomel, six grains, followed by a saline purge and hot-water enema. Give ten grains of bicarbonate of soda three times a day. Give cooling drinks, such as fruit salt. Make the skin act. Apply hot fomentations to the back and mustard leaves to the pit of the stomach.

Burn or thoroughly disinfect all discharges and clothing; boil all cups, spoons, etc., which the patient has used.

Filariasis.

The filaria, which is a small blood worm, is conveyed by the bite of the mosquito of the genus Culex. It may help in producing a series of symptoms known as elephantiasis, and hence the importance of protection from the bites of mosquitoes.

Plague.—(Bubonic plague.)

A disease characterised by high fever and delirium, accompanied by the formation of glandular swellings in armpit, groin, or neck.

Causes—Bubonic plague is contagious from man to man; the infection may be carried by rats, soil, or water, or the germs may gain an entrance to the body through a cut or abrasion. The chief predisposing causes of plague are filth, overcrowding, bad ventilation and drainage, and impure water supply. The disease attacks people of all ages, but is less liable to attack white than coloured people.

Incubation period, two to seven days.

Symptoms.—The disease commences with lassitude, headache, giddiness, and shivering. The temperature rapidly runs up to 102° F., 104° F. or even higher, and is usually accompanied with restlessness, sickness, and diarrhœa. The tongue becomes dry and brown, and there may be delirium; the pulse fails and the extremities become cold.

After one or two days the glandular swellings commence in the armpits or groin. The glands may reach the size of a hen's egg, and if the patient lives, break down and discharge matter. Boils and carbuncles may also occur during the progress of the disease.

Death usually takes place before the sixth day, but in cases that recover convalescence usually commences by the tenth day.

In one form of plague, inflammation of the lungs is the most characteristic sign.

Treatment.—If possible, inject plague antitoxin; if this is not available give quinine, and treat the symptoms, such as thirst and high temperature, by antipyrine, salicylate of soda, etc. Give stimulants for collapse, astringents or opium for diarrhœa, and sedatives for vomiting.

Rheumatism.

This is a disease which frequently follows exposure to damp and cold, and is on that account not uncommon in the tropics. It is often hereditary. After one attack, rheumatism is always liable to recur in the same individual, and on this account it is necessary that persons liable to the disease should use special precautions.

Acute rheumatism. Symptoms.—This begins by a shivering fit, with rise of temperature and general sickness, and the joints, usually wrists, ankles, or knees, become painful, tender, and afterwards swollen. It resembles other feverish conditions in the rapid pulse and breathing, the constipation, scanty and high-coloured urine, etc., but it differs from most of them in the presence of a profuse and sour-smelling perspiration, resembling the odour of butter-milk.

Treatment.—The best remedy for acute rheumatism is salicylate of soda, of which fifteen grains should be given every six hours. The joints should, at the same time, be kept wrapped up in cotton wool, covered with oiled silk and a flannel bandage. This treatment will nearly always correct the acute symptoms in two or three days. If the symptoms subside sooner, the quantity of the salicylate should be diminished; if there is delirium, the dose must be lessened at once, for many people are very susceptible to salicylates and are easily affected by them, the delirium being characteristic. Dover's powder may be given to relieve pain and to secure sleep.

Chronic rheumatism.—In this disease there is chronic pain and tenderness of the joints, without fever.

Treatment.—Bicarbonate of potash and salicylate of soda, each in five-grain doses, should be given every eight hours. Painful joints may be painted with tincture of iodine, rubbed with turpentine liniment, or bathed with hot water. The bowels should be kept well open, and alcohol and much meat avoided.

Faintness.

Bending the head firmly down between the knees is the best and most rapid way of dealing with an attack. Another method is to lie the patient on a couch, loosen any clothing which is round the neck, bathe his face and arms with cold water, and fan him vigorously. Give three grains of carbonate of ammonia in an ounce of water. Alcohol may be given if the pulse is very weak, but sal volatile or carbonate of ammonia is more efficacious.

Sunstroke.

This is a dangerous affection, but is not so common in the tropics as is generally supposed. It may be caused either by direct exposure to the sun, or exposure to a superheated atmosphere. It must be guarded against when the surrounding temperature exceeds that of the normal temperature of the body, *i.e.*, 98·4° F. During the nine years the author spent in Nigeria very few cases were noted.

Symptoms.—In a large proportion of cases warning symptoms are present, such as giddiness, headache, sickness, heat and dryness of the skin, bloodshot eyes, and frequent desire to pass water. In these cases the attack can usually be warded off by prompt treatment. Sometimes, however, the patient suddenly falls down insensible.

Treatment.—The patient should at once be drenched all over, especially on the head and along the spine, with cold water, and this treatment should be kept up for a considerable time. Six grains of calomel should be placed at the back of the tongue, and, if the patient is conscious, washed down with a saline such as fruit salts; in very urgent cases a turpentine or hot-water enema should be given as quickly as possible. The patient can then be put to bed, with cold wet cloths wrapped round the head and adjusted along the spine; these should be frequently changed and wetted, the cooler the water the better, and this treatment should be persisted in till the patient is roused and sensible. When procurable, an ice-bag may be applied to the head.

Concussion of the Brain.

This term is applied to the partial suspension of the functions of the brain, produced by the severe shaking of its substance by a fall or blow.

Symptoms.—At first the patient lies in an unconscious condition, skin cold and clammy, pulse and breathing very feeble, and temperature extremely low; he can be slightly roused by shouting; he cries out if he is moved, or when painful applications are made, but quickly relapses into insensibility. The stage of unconsciousness may pass off almost at once, it may be prolonged for hours or days, or the patient may never recover from it. The second stage—that of reaction—is marked by returning consciousness and frequently by vomiting, the skin becomes warm, and gradually the patient recovers; on the other hand, inflammation of the brain may set in, or he may again become unconscious and die.

Treatment.—Keep patient perfectly quiet in bed, in a darkened room, give a milk diet, and if he is much excited, apply cold cloths or an ice-bag to the head. If there is much prostration apply a hot-water bottle, and restore the circulation by rubbing the limbs. When reaction sets in, give five grains of calomel.

Stimulants should be avoided in cases of concussion of the brain, unless the collapse is very alarming (when ammonia should be given), as they tend to cause too violent reaction, which might be followed by inflammation of the brain and its coverings.

Stroke, or Apoplexy.

This disease is caused by the rupture or blocking up of one of the blood-vessels in the brain.

Symptoms.—The person attacked falls down suddenly, and is unable to move one or more of his limbs. He may be quite insensible, or soon become so, or perhaps he is unable to talk. The mouth may be drawn to one side, and the tongue, when protruded, be pushed to the right or left. The condition is serious.

Treatment.—Tight clothing must be removed. Six grains of calomel powder should be placed on the back of the tongue, and the patient kept lying on his back with the head slightly raised. Cold should be applied to the head and a hot-water bottle to the feet, the room darkened, and absolute quiet observed. An enema of hot water may be given, and while the patient is insensible the lips should be moistened only with water. Food may be given by the bowel on the second or third day. Stimulants are absolutely forbidden. If patient gets over the attack he ought to be sent home.

Note.—It must not be forgotten that many of the above symptoms might be caused by injury or poison.

Epilepsy.

Symptoms.—This is the most common form of fits. There are three stages. 1*st stage*—The patient falls down completely unconscious and without warning, the face is pale, the limbs become stiff and rigid, and the breathing ceases. 2*nd stage*—Convulsive movements take place, the tongue being often bitten, the breathing becomes heavy and laboured, and the motions may be passed unconsciously. 3*rd stage*—A confused mental condition, sometimes acute mania, usually sleepy for some time. In some cases the fit may only last a few minutes. They almost always recur.

Treatment.—During the attack nothing can be done beyond loosening all tight clothing, and gently preventing the sufferer from injuring himself in his struggles. It is especially important to keep the teeth apart with a piece of cork or rubber to prevent the tongue from being bitten.

Bromide of potassium (ten to twenty grains) may be given three times a day as a preventive.

The Infectious Fevers.

In discussing the following fevers, the term 'incubation period' frequently occurs, and some explanation of this term is necessary. By incubation period is meant the time which elapses from the date of exposure to infection to the first symptoms of illness—not the appearance of the rash. The date of eruption of the rash is calculated from *the first symptom of illness.* Some of the fevers already described might also be included under this heading.

Scarlet Fever, or Scarlatina.

This fever is very uncommon in hot climates.

Incubation period, three to eight days. Rash appears second day.

Symptoms.—The rash consists of numerous minute red spots, evenly distributed all over the surface of the body, upon a general rose-red blush. The area immediately surrounding the mouth is not affected. Shivering fits, sickness, high temperature, sore throat, headache, pains in the back. The sickness is very characteristic in children. After the rash has subsided, *peeling* of the skin takes place, beginning on the face as a very fine powdery deposit. The process then spreads to other parts of the body, the last places to peel being the palms of the hands, the soles of the feet, and between the fingers and toes. The peeling process takes from four to six weeks, and the patient is infectious during the whole of this period.

Treatment.—Isolation. Bed in an airy room. Light food. If the fever is high (104° F.), the body may be sponged with tepid water. If the throat is very inflamed, an antiseptic gargle should be used. As the course of infectious fevers cannot be cut short, the chief aim of treatment should be to avert complications and the return of distressing symptoms—over-treatment with drugs must be guarded against.

To prevent the spread of infection from the peeling surface, the body should be rubbed over with boracic ointment, or carbolic or olive oil.

Complications.—As scarlet fever may be followed by heart, kidney, or ear trouble, the patient should not be allowed to get up until at least a week after the acute symptoms have subsided.

Measles.

Incubation period, ten days. Rash occurs on the fourth day.

Rash.—Pink spots, round or irregular, slightly raised above the surface, tending to run together in patches over the body, leaving the unaffected skin between them clear. In the early stages often best marked on the face and behind the ears.

Symptoms.—Fever, catarrh, congested eyes, running from the nose, sickness and cough.

Treatment.—Similar to that of scarlet fever.

Complications.—Measles may be followed by pneumonia.

Smallpox.

Very prevalent in the tropics, hence the importance of revaccination before going abroad.

Incubation period, twelve days. Rash appears third day.

Rash.—Before the appearance of the typical rash there are occasionally earlier rashes, viz., a diffuse blush covering the whole body, resembling scarlet fever, or a dark purple rash of effused blood beneath the skin of the lower part of the belly, or occasionally in the armpit. The smallpox rash proper consists of small red raised spots which first appear on the face, forehead, and scalp, subsequently coming out over the rest of the body, commencing at the top and working downwards. These spots become prominent, and have a characteristic "shotty" feeling under the finger. On the third day after their appearance a small bleb forms in the centre of each spot; it is transparent at first, but subsequently becomes yellowish, from the formation of matter in its interior. The centre becomes depressed on the sixth day, then the bleb breaks down, and discharges matter. Two or three days later the spot begins to dry up, and ultimately heals under a scab. During the period of most active inflammation the face may be very swollen and sodden.

The spots and pustules are not confined to the skin, but may occur on the roof of the mouth and in the throat.

Treatment.—Isolation, similar to that of scarlet fever. The body may be sponged and vaseline applied if there is much itching. The eyelids and eyes should be frequently washed with weak boric acid solution.

Chicken-Pox.

Incubation period, a fortnight to three weeks. Rash appears first day.

Rash.—Pink spots, upon which blebs form after twelve to twenty-four hours. The blebs are at first transparent, but subsequently become yellowish, and after two to three days shrivel and separate, leaving a pink scar.

The symptoms are usually very mild, perhaps only slight fever, and possibly headache. The appearance of the rash is often the first symptom.

Treatment.—Isolation, and light diet. Bed may not be necessary.

Diphtheria.

Diphtheria is an acute infectious disease, the essential feature of which is a peculiar inflammation of the lining membrane of the mouth, nose, throat, and windpipe, characterised by the formation of a membrane upon the inflamed surface.

Causes.—It may be contracted from some person suffering from the disease, or from infected milk, etc.

As the disease is a very grave one, and skilled treatment is often an absolute necessity, measures should be immediately taken to summon medical assistance on the first appearance of diphtheria, or the patient should be sent as speedily as possible to a place where medical aid is likely to be obtained; for if the breathing becomes so difficult that the patient gets blue in the face, an operation for opening the windpipe will be necessary.

Incubation period, two to six days or even longer.

Symptoms.—Headache, discomfort, loss of appetite, sore throat, and sickness, with swelling of the glands at the angle of the jaw. On examination the palate and tonsils are seen to be swollen, with a white deposit of membrane upon the surface. The membrane may be thick and tough, and if stripped off will leave numerous small bleeding points.

The temperature may run up, and is irregular in type. The pulse is rapid and feeble, and the bodily strength is quickly lost.

If the nose be affected there is copious discharge from the nostrils, with difficulty of breathing and much discomfort. If the windpipe is affected the voice will become hoarse or absent, and there will be greater difficulty in breathing, accompanied by a loud crowing noise.

Diphtheria may be accompanied by cough and pneumonia.

Treatment.—Isolation. Bed. If diphtheria antitoxin is obtainable, it should be administered at the earliest opportunity, 1500 "units" being injected under the skin as the first dose, but this should only be carried out by a doctor.

Nourishing foods and stimulants should be given frequently in small quantities. The throat should be thoroughly and frequently washed out with chinosol (1 in 1000) or other antiseptic lotion. If the difficulty in breathing is marked, warm baths should be given at intervals of about four hours. A steam kettle should be placed near the bed. The expulsion of the membrane may often be aided and great relief afforded by the administration of emetics, such as ipecacuanha.

Complications.—Diphtheria may be followed by paralysis of the windpipe with loss of voice, or paralysis of other parts of the body, therefore great care should be taken not to allow convalescents to get up too soon, no matter how well they may appear.

Coryza, or Cold in the Head.

When a cold is confined to the head it can usually be cut short by retiring to bed early, taking a ten-grain dose of Dover's powder, followed by hot drinks to encourage the perspiration which the action of this drug produces, together with the use of as many additional bed-clothes as can be borne. Care must be taken to avoid chill on the following morning. In tropical regions five grains of quinine should be added to the dose of Dover's powder.

Hay Fever.

This is a very severe catarrh which attacks certain individuals yearly, when grasses and other plants are flowering. It is most probably due to the irritation of the nose by pollen grains in the air.

Treatment.—Exposure to the irritating substances which are known to provoke an attack should be avoided.

The nose may be syringed out with a lotion containing boric acid and bicarbonate of soda (five grains of each in four ounces of water), or one containing bicarbonate of soda and salt (five grains of each in four ounces of water), to which has been added two to four drops of carbolic acid. The interior of the nostrils may be anointed with vaseline. Menthol snuff is of great value.

Laryngitis, or Inflammation of the Upper Part of the Windpipe.

The organ of the voice is called the "larynx."

When the windpipe is affected it is somewhat tender on pressure, there is hoarseness, cough, and pain in swallowing. Treatment similar to that for cold in the head may be adopted; in addition, the upper part of the front of the throat should be kept well poulticed for a day or two, and then wrapped up in cotton wool for some days longer. A piece of mustard leaf covered with six layers of a handkerchief and secured by a bandage, can usually be borne for a considerable time, and is often more efficacious than the poultice; when the smarting is great the mustard leaf should be removed and the tender part smeared with oil or vaseline.

Inhalations of steam are of use in promoting expectoration. Carbonate of ammonia, three grains, or bicarbonate of potash, five grains, with half to two grains of ipecacuanha, will help to promote secretion from the affected part.

Quinsy, or Inflammation of the Tonsils.

Apply poultices to the neck. Gargle with a hot, weak solution of permanganate of potash, or a solution of chinosol (1 in 2000), at least every hour. Administer quinine and iron as a tonic. Keep the bowels well open.

Ulceration of the Throat.

Gargle with a weak antiseptic solution such as permanganate of potash or chinosol. An astringent gargle may be made by dissolving five grains or more of tannin in two ounces of hot water. Sulphate of iron can be used for the same purpose, two grains or more to an ounce of water. If there is much pain, apply poultices to throat.

If due to syphilis give one grain of calomel and three grains of iodide of potassium, twice a day, in addition to the local treatment.

Bronchitis, or Inflammation of the Branches of the Windpipe.

Symptoms.—When bronchitis exists, there is a good deal of coughing—at first dry, but afterwards accompanied by frothy expectoration—with a sensation of rawness and tenderness at the upper part of the breastbone.

Treatment.—In the early stages of this condition, opium in some form or other will be found beneficial, and will often cut short an attack; for this purpose, ten grains of Dover's powder, or fifteen to twenty minims of chlorodyne, may be given every eight hours for twenty-four hours, and then be gradually diminished.

If the breathing is difficult, poultices should be applied to the chest, and ipecacuanha, half to two grains, and ammonia, should be given three times a day.

Inhalation of steam often gives great relief; and the effect is much improved if thirty drops of Friar's balsam are added to a pint of hot water.

In tropical climates even an ordinary feverish cold very often tends to become malarial in character, therefore the use of quinine, in addition to the other treatment, is usually desirable, and five grains may be given twice a day.

Pleurisy, or Inflammation of the Membrane Covering the Lung.

This is more a disease of cold climates, and is usually the result of chill following severe exertion.

Symptoms.—Pleurisy is accompanied by less fever and general sickness than pneumonia; its characteristic symptom is the "stitch in the side," which always accompanies it. There is also a short, dry cough, without expectoration, which the patient tries to restrain, as it "catches" in the side, and causes acute pain. For the same reason the breathing is shallow, as any attempt to draw a deep breath causes extreme suffering.

Treatment.—The patient suffers greatly, therefore in the early stages treatment must be directed to the pain. If leeches are procurable, the application of half-a-dozen to the painful region of the chest is advisable. Mustard leaves or poultices should be applied over the part, or it may be painted with tincture of iodine; opium may be given to relieve the acute pain, in the form of Dover's powder, fifteen grains three times a day. Five grains of quinine should be given twice a day.

Pneumonia, or Inflammation of the Lungs.

Symptoms.—This usually begins with a severe attack of shivering; the temperature rises rapidly, the pulse and breathing are greatly quickened, and the patient is completely prostrated. The face is flushed, the skin feels hot and dry, and there is a short cough, dry at first, but afterwards accompanied by expectoration of a moderate quantity of slimy, rust-coloured, blood-stained, and almost frothless matter. Usually there is pain on the affected side, which in most cases is the right side of the chest, above the liver.

Treatment.—A patient attacked with pneumonia should take to bed at once. The affected side should be surrounded with a *large* poultice. Five grains of quinine should be given every eight hours. If the heart's action is weak, give some preparation of ammonia, as a stimulant, and administer alcohol, up to half an ounce, every two hours. Opium should only be given to calm the patient, as large doses do harm by checking free expectoration; if there is much distress, then ten or fifteen grains of Dover's powder may be given.

An ice poultice applied to the chest will give great relief by lowering the temperature and diminishing the pain (*see* page 256).

Simple Ophthalmia.

Simple ophthalmia, conjunctivitis, or inflammation of the membrane covering the eye and the inner side of the eyelids, is usually due to cold or dust.

Symptoms.—The affected eye is bloodshot and painful, waters freely, and cannot bear a bright light; there is a feeling of grittiness, as if the trouble were due to something between the eye and eyelid.

Treatment.—The eye should be carefully washed, the eyelids being opened and clean water allowed to run over them and over the eye; any particles of dust must be removed—for this purpose a small clean camel-hair brush will be found useful. A lotion should be made consisting of six grains of boric acid, or two grains of sulphate of zinc, to an ounce of water, and ten drops or more of this should be dropped on to the eyeball six or eight times a day. A pad of lint soaked in clean, cold water should be applied over the eye. Sometimes hot fomentations give relief.

Purulent Ophthalmia.

This is a more serious inflammation, and is caused by some poison, *e.g.*, germs carried by flies, or by the fingers from unhealthy sores and discharges.

Symptoms.—The symptoms of simple ophthalmia are present, but are all intensified, the eyelids are swollen and the eyeballs red, there is a discharge of yellow matter or pus, and the patient feels ill. There is great danger of the affected eye infecting the sound one, therefore warn patient not to touch the sound eye for fear of infecting it. There is also great danger lest the attendant's own eyes should become infected.

Treatment.—The patient should be kept in bed and the eyes should be shielded from bright light. Protect the sound eye (especially when the affected one is being washed) by placing a pad of wool or lint over it, kept in its place by strips of strapping so as to effectually close the eye and prevent infection. Thoroughly wash out the space between the eyelids and the eye, and remove any matter or foreign body which may be found.

When the inflamed surfaces are clean, wash them very thoroughly with a solution of corrosive sublimate, 1 in 5000, and finally smear a little vaseline along the edges of the lids, to prevent them sticking together. This treatment must be repeated as frequently as possible. Once a day the inflamed surfaces may be brushed over with a solution of nitrate of silver, 10 grains to the ounce, applied with a camel's-hair brush, followed immediately by the application of a few drops of common salt solution.

Hot fomentations may give relief. When this is so, the eye should be kept covered with a pad of moist lint, which must be changed frequently.

Iritis, or Inflammation of the Eyeball Itself.

Symptoms.—In this there is pain, the vision is dimmed, and the transparent part of the eye is found to be cloudy. Skilled assistance is necessary.

Treatment.—Apply hot fomentations and boric acid lotion; leeches or a blister to the temples are of service, and the pupil, which is contracted, should be dilated by dropping two or three drops of a one per cent. solution of atropine on to the eye twice a day or oftener till it is well dilated; only enough drops should be applied afterwards to prevent the pupil from contracting. Keep the eye covered with a pad of lint soaked in cold water. The bowels should be kept well open, and one grain of calomel may be given three times a day for a week, or longer if it does not cause a coppery taste in the mouth, with tenderness of the gums and excessive flow of the saliva. As the inflammation subsides a shade may replace the pad.

Night blindness and snow blindness are due to exposure to the glare either of the sun or of the snow. To avoid these complaints tinted glasses should be worn. Travellers in snowy regions should be provided with smoked glasses; if these get broken or lost, some opaque substance may be smeared over the surface of an ordinary pair, leaving a narrow horizontal slit of clear glass—in the Esquimaux fashion, as shown in the accompanying illustration. On snow it must be remembered that

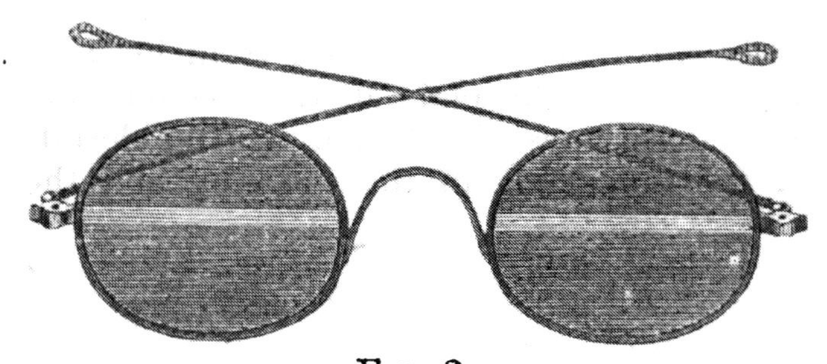

FIG. 2.

the perforated wire gauze sides are essential for protection from the reflected rays of the sun. Elastic may be substituted with advantage for the ordinary metallic attachments, between the glasses as well as around the head.

Sleeplessness.

This is common in those broken down by malaria, dysentery or debilitating diseases; it is also induced by the irritation caused by mosquitoes and other pests, such as the itch parasite. It is at times due to errors of diet, prickly heat, mental worry and exhaustion, abuse of tea and coffee, coldness of the feet and indigestion, and as most acute diseases are worse at night, sleeplessness is very common amongst sick people.

Treatment.—As far as possible remove the cause; kneading the feet and legs, or the application of a hot-water bottle to the feet, will often be of service. A cup of hot milk or soup should be taken at bed-time, and again on waking in the night. The bowels must be regulated and alcohol taken very sparingly.

The taking of sedative drugs should be avoided as much as possible. The least harmful of these is sulphonal in doses of from twenty to thirty grains, given several hours before bedtime, or bromide of potassium in twenty or thirty-grain doses given at bedtime. A warm bath at night often acts more satisfactorily than any other remedy. Only very rarely should chloral, chlorodyne, or opium be resorted to.

Regular habits and plenty of work are potent factors in the production of that healthy condition which predisposes to natural refreshing sleep.

The unfortunate man who has no hard work to do, who is without even a hobby to occupy him, and has no interest in life but the torpid condition of his liver, is a constant sufferer from insomnia. He should be purged and put on a light plain diet. Alcohol must be forbidden and plenty of exercise must be taken. A tumblerful of hot water is to be taken the first thing in the morning and the last thing at night.

Sea-Sickness.

Take a saline aperient on the day before embarking, and a light plain meal at least three hours before going on board. A cup of good tea or black coffee soon after starting is often of use.

Those liable to sea-sickness should go to bed directly they get on board: the head should be kept low and the room darkened. A mustard leaf applied to the pit of the stomach is of value in diminishing the tendency to vomit. An abdominal belt is useful from the gentle support it gives. A hot-water bottle may be applied to the feet.

A mixture containing fifteen grains of bromide of soda and five grains of antipyrine to one ounce of water is often of great value. The first dose should be given immediately the patient is in bed, and may be repeated every six hours if required.

Whitla states that the best of all remedies is bromide of ammonium, It should be given in twenty-grain doses for a day or two before embarking. Morphia may be found necessary: a third of a grain may be injected under the skin of adults, but it should on no account be given to children.

I have found that three or four drops of chloroform, dropped on to loaf sugar and sucked, often prevents vomiting.

Those who are vomiting severely should take plenty of hot water or milk, so as to prevent them from straining on an empty stomach.

Prickly Heat.

Prickly heat is frequent in the tropics. It is due to free sweating, and causes intense heat and itching.

The diet should be light, very little animal food, and no alcohol. The skin should be kept clean, cool, and dry, and light linen or cotton garments worn next the skin instead of flannel.

The annoying heat and itching are relieved by the application of carbolic solution to the skin (about a tablespoonful of carbolic acid to two pints of water). A dusting powder of starch, arrowroot, or oxide of zinc often does good.

A lotion made by squeezing three or four limes into a pint of water may be applied, but this often causes for the moment a good deal of smarting. Goulard lotion often does good in many cases. It may be necessary to give twenty grains of bromide of potassium at night to procure rest. Bicarbonate of soda should be given three times a day in ten-grain doses. A good saline purge should always be administered at the beginning of an attack.

Chilblains and Frostbite.

Chilblains are usually found on the fingers or toes—after exposure to severe cold—especially when tight gloves or boots have been worn. Certainly the best way to promote the formation of chilblains is to toast the semi-frozen fingers or toes at a fire or stove, before the circulation has been re-established.

When chilblains are threatened, the part should be well rubbed with snow, or with camphorated spirit. Sponging with hot vinegar is very effective. Chilblains are checked in the early stages by painting with tincture of iodine. Ulcerated chilblains should be dressed with boric ointment spread on lint.

Prolonged exposure to intense cold leads to development of frostbite. If the case is a bad one, or injudiciously treated, gangrene or death of the part always follows; if this is extensive, amputation may be necessary.

Frostbite should be treated first by vigorous friction with snow or pounded ice. The affected parts should then be well wrapped with cloths wet with cold water. It is extremely dangerous to bring the frozen parts near a fire. Afterwards, the part should be wrapped in cotton-wool.

Scurvy.

Scurvy is a disease generally caused by the consumption of salted food, and deprivation of fresh vegetables for a considerable period.

Symptoms.—Loss of colour, weakness, drowsiness, languor, and apathy, with possibly pains in the limbs. After a time small spots appear all over the body, caused by the effusion of blood beneath the skin, the gums become spongy and tender, and the teeth are loosened. If the condition persists the patient becomes sallow, bloated, breathless on exertion, and totally unfit for work, and may have fainting fits. If not relieved, death may supervene. Scurvy is often associated in its later stages with dysentery.

Treatment.—Fresh vegetable food, potatoes, green vegetables or salad. Fresh meat once a day. Lime juice, jam and fruit are also very valuable as routine preventatives. It will often be found advisable to adopt, in some measure, the dietary used by the natives.

Boils.

Boils are very common in the tropics. They should be poulticed or fomented frequently. When there is a yellow head they may be opened, but as a rule they should be allowed to burst; after the boil has discharged, some healing ointment, *e.g.*, boric or iodoform ointment, should be applied.

Plain, good living is essential, and the patient's general health should be attended to with tonics, lime juice, and green vegetables. If the sufferer is found to be taking too much alcohol, it should be stopped altogether. Sulphide of calcium pills, one grain in each, should be given three times a day to prevent boils forming.

Carbuncles.

A carbuncle is of the same nature as a boil, but is a more serious complaint, and it is distinguished from a boil by the fact that it opens by several mouths.

To promote ripening of the carbuncle and the separation of the core or dead piece of tissue, poultices and hot fomentations should be freely applied, and if the core can be seen it should be removed, if possible, by a pair of forceps.

To facilitate removal of the core it is sometimes advisable to cut through the skin separating the openings; there may be some slight bleeding, but it will cease after a short time.

Opium may be given to relieve acute suffering. The bowels should be kept well open. A generous diet must be given; tonics of iron, quinine and arsenic are needed, and alcohol may prove necessary.

Carbuncles rarely appear except in people much broken down in health, and their presence is an indication that the sufferer should return to his own country and seek skilled advice.

Ulcers.

Ulcers are often very troublesome to the traveller, as the healing process is frequently retarded by exposure, dirt and dust, and the chafing of clothes. Want of sufficient animal food greatly favours the spread of large ulcers.

A simple dressing of iodoform ointment, or ointment of boric acid, is effective when the ulcer can be protected, and rest can be given to the affected part. When the ulcer is deep and large it may be stimulated to heal by the application of lint or cotton-wool, moistened with carbolic oil, carbolic lotion, or other antiseptics.

Large, unhealthy ulcers should be well bathed with some antiseptic such as carbolic acid, one in sixty of water, or a solution of chinosol (1 in 600); and after the sore has been cleansed it may be lightly dusted with iodoform and then covered with a piece of lint moistened with carbolic oil or smeared with iodoform ointment and supported by an evenly applied bandage. Boric ointment, lano-creoline, izal ointment or other antiseptic dressing may take the place of iodoform ointment.

If the ulcers are very irregular in shape and appear to be due to the later stages of syphilis, iodide of potassium, from 5 to 10 grains three times a day, may be given.

Iron and quinine tonics should be given.

Bedsore.

When a patient is confined to his bed for a long time it is necessary to support the lower part of the back on something soft, such as a pad of wool, or an extra pillow, in order to prevent the formation of a bedsore; pillows should also be placed under the hips and heels.

To harden the skin it is well to rub with oil or white of an egg three parts, and spirits one part; zinc ointment is also useful for this purpose.

Besides pressure, the most frequent cause of bedsores is constant moisture from the passage of urine and motions and consequent damping of the sheets. Great care must be taken to thoroughly dry the back after any evacuation; the lower part of the back should be dusted with a powder of boric acid and zinc oxide.

If a bedsore is present the patient should lie on a circular pad with a hole in the middle, to take pressure off the bedsore. The sore should be thoroughly cleaned twice a day with some antiseptic lotion, such as chinosol or permanganate, and subsequently dressed with zinc ointment or vaseline.

If the bedsore is on the back, the patient should lie if possible on each side alternately, to relieve the pressure.

Parasites.

Mosquitoes and Midges.—These and other annoying insects must be warded off by the use of netting, as already described. The application of carbolic, castor, oil of lavender, or eucalyptus oil or kerosene helps to keep them off, whilst the use of antiseptic soaps, such as Izal soap, Wright's coal tar soap or carbolic soap may also be of service, but none of these measures must be relied upon.

The application of ammonia to mosquito bites diminishes the irritation.

Lice on the head or body can be destroyed by using white precipitate ointment, mercurial ointment, or carbolic oil. Care must be used in applying strong ointments; at first they should be used sparingly and well diluted with oil or vaseline. Much time will be saved by shaving off or cutting short the hair. The whole body should be frequently washed with carbolic soap and hot water. Clothing should be boiled or destroyed.

Ringworm of the head is a disease due to a fungus. The part for some distance round the affected area should be shaved, and any stumps of hairs pulled out. The skin should be lightly painted with tincture of iodine or strong solution of carbolic acid (one in ten of water); or sulphur, white precipitate, or mercurial ointment may be gently rubbed in.

Dhobie itch.—This complaint, which is very common in the tropics, attacks the groins and the upper part of the thighs, and it is frequently found in the armpits. It is due to a fungus similar to that of ringworm, and is aggravated by free sweating and the rubbing together of adjacent parts of the body. The parts actually affected will be seen to be slightly raised and "scurfy" and somewhat reddened. It causes great itching, and at times, on account of the irritated skin being inflamed, much pain is associated with it.

Treatment.—When the parts are inflamed they should be soothed by the application of Goulard water or Wright's liquor carbonis detergens (one tablespoonful to two pints of water), zinc ointment, starch, or zinc dusting powder. Inflamed surfaces should be kept apart by means of a piece of lint smeared with zinc ointment placed between them. A few days' complete rest, in cool linen clothing, will give great relief.

When the inflammation has subsided, the affected parts can be rubbed with ichthyol ointment, araroba ointment, white precipitate ointment, or sulphur ointment. Such remedies as tincture of iodine, araroba powder or white precipitate ointment should be used at first very sparingly, otherwise they are liable to cause pain and inflammation.

Scabies, or itch, is due to an animal parasite, the female of which burrows into the skin to lay her eggs. The disease is commonest between the fingers, but it may spread up the arms to the body, and down the legs. The parasite causes an eruption and produces very great itching, especially at night.

Treatment.—Wash the skin well with soap and hot water, then freely rub in sulphur ointment. A clean linen suit should be worn night and day, and the ointment should be applied every night for three or four nights, or till the irritation ceases. The clothing should then be well boiled or destroyed.

Kraw Kraw is the name applied to a contagious disease which is extremely common in certain parts of the tropics, especially Africa; it is accompanied with great irritation and itching of the skin. Frequently the disease first appears as a patch, not unlike ringworm, on the wrist, arm, or thigh; the patch, which is pimply, becomes in parts pustular, *i.e.*, it contains matter. Other patches appear and spread, the patient being re-infected either by his clothes or by scratching himself.

Treatment.—The parts must be well washed, and then araroba ointment, white precipitate ointment, or sulphur ointment should be applied. Araroba ointment is the best of all applications, but is liable to cause some congestion and irritation, and even inflammation of the skin; therefore only a little should be applied at a time, or, better still, the ointment should be diluted by mixing it with four times its bulk of vaseline.

The disease is uncommon amongst white men or superior coloured men, and when these are attacked care must be taken in the use of strong ointments. Amongst the labourers the disease is very common; these men like something with a sting in it, and would feel hurt if they were put off with milder preparations than croton oil for internal, and araroba ointment for external use.

Where suitable ointments are not available, then carbolic soap, carbolic oil, one in twenty, or lotion, one in forty, or tincture of iodine may be used.

The Jigger or Sandflea.—The female penetrates the skin, usually of the feet, especially near the toe-nails. As the insect enlarges it causes considerable pain and irritation.

Prevention.—Keep the floors of the houses very clean, and never walk about with the feet bare.

The *treatment* is to turn out the little bladder-like body with a clean needle, and dress the sore with an antiseptic ointment. Natives are usually very expert in removing these pests.

Ticks cause considerable trouble and may convey infection; they are sometimes found in the nostrils. They should be removed as quickly as possible, and they will come away easily if they are well covered with oil.

Leeches abound in the long grass of certain tropical districts, notably of India and Ceylon. They attach themselves to the skin and have often been known to creep into the nostrils, where their bites cause considerable loss of blood. They are best removed by the application of salt, or by the injection of salt and water.

Fleas and Bugs may be kept at a respectful distance by the use of "Keating's insect powder."

Keating's powder is slow in its effects, and therefore, if possible, should be shaken over the sleeping bag or blankets some hours before bedtime. If not, the pests will struggle through it and find renewed vigour on the sleeper. It is best in very bad quarters to rub the powder on the skin as well as to dust it over the bed. It will not kill a full-grown bug under an hour, but it is extraordinarily effective with fleas. Block up all except one small aperture, and put the powder freely about that. In a few minutes fleas will be lying stupefied all over the floor, and must be swept up and destroyed, or they will revive.

Worms.

Worms are introduced into the system chiefly by means of dirty water or imperfectly cooked food.

Of the worms which live in the bowels the most important are the following:—

Tape Worms.—These worms may measure many feet in length; and their presence in the body can only be certainly known by the appearance of some of the segments or portions of the worm in the motions; although hunger and dyspepsia may be complained of.

Treatment.—Administer a good aperient overnight so as to empty the bowels; after the aperient nothing should be given by the mouth for eight hours, then give sixty to one hundred and twenty drops of the liquid extract of male fern in one ounce of thin gruel, milk, or gum water, and follow this up in four hours by a good meal, and an aperient to remove the worm, which should now be dead.

When the extract of male fern is not available, one tablespoonful of the oil of turpentine may be used in its place.

If later on fresh segments appear in the motions, then the treatment must be repeated.

Round Worm.—The round worm resembles the garden worm and is several inches in length; it may be observed in the vomit but more commonly is seen in the motions. The symptoms are similar to those caused by the tape worm and the treatment is the same, except that, instead of the male fern, two to five grains of santonin should be given in a little milk; and the treatment repeated every other day for a week.

Threadworm. — The threadworm is a small round worm usually measuring less than half an inch in length; it inhabits the lower end of the bowel and causes great heat and itching about the outlet, especially at night.

Treatment.—Wash out the lower bowel and inject into it about a third of a pint of tea, or a similar quantity of water containing one teaspoonful of salt, tannin, or alum; then apply a little mercurial ointment around the outlet to diminish irritation. This should be done every third day till the worms have disappeared from the motions.

The Guinea Worm.—The guinea worm often measures several feet in length; it chiefly causes trouble in the feet, ankles, and legs, where in order to obtain an exit from the body it penetrates the skin, causing a small ulcer at its point of exit.

Usually the presence of the guinea worm is attended with inflammation and the formation of matter. When the worm can be seen at the base of the little ulcer, it may be secured to a piece of match and a small portion may be wound on to the match daily. If attempts are made to forcibly draw it out, it will probably break and violent inflammation will result. During the time that the worm is being wound out, the part should be kept very clean and an antiseptic ointment applied.

Other Worms producing Disease.—There are various other minute worms which may exist as parasites in the human body. One of these, known as the *Ankylostomum duodenale*, produces a form of anæmia which is most common in Egypt. It is difficult, however, to recognise the diseases due to this and other parasites without the use of a microscope, so that it is of no use to give details of treatment in these cases; but the importance of cleanliness of habits, and care with reference to drinking water, should be insisted, as most of these diseases are acquired in some way through insanitary conditions.

Rupture or Hernia.

A rupture or hernia is a protrusion of some portion of the bowels under the skin, and is usually found in the groin. It is generally reducible, *i.e.*, it can be pushed back into the belly. It reappears when the pressure is removed, especially if the patient coughs or strains. When reduced, a properly-fitting truss should be applied and worn during the day, it can be taken off at night, after lying down, but should be re-adjusted in the morning, whilst the patient is still in bed. No patient should go abroad without having an operation for the cure of the hernia.

The great danger of any rupture is that it may become irreducible—a condition which is very likely to be followed by constriction or "strangulation" and subsequent death of the ruptured part of the bowel. If unrelieved, this constricted condition is always fatal. The existence of strangulation is known by local pain and tenderness, development of severe colicky pains in the belly (especially about the navel), absolute constipation, vomiting, hiccough, and symptoms of collapse. When this condition is observed, the patient's hips should be raised by supporting them with pillows, and the tumour should be only very gently kneaded with the view of getting back the protruded bowel. The treatment is considerably aided by immersing the patient in a warm bath, and giving about twenty drops of laudanum or chlorodyne. Ice placed round the swelling for half an hour or so is often very effective. If these means fail, surgical aid is absolutely necessary. *Purgatives should not be given.*

Retention of the Urine.

Retention or inability to pass the water may be caused by stricture, injury, shock, spasm, inflammation of some part of the passage, the effects of drinking, or by chill.

Symptoms.—The bladder is unable to expel its contents and it gets fuller and fuller; it can be felt in its distended condition as a painful, soft swelling in the lower part of the belly, below the navel, underneath the skin and muscles. There may be fever, great pain and constant desire to pass water, with inability to do so. When the bladder becomes greatly distended, there is usually slight dribbling of water, which is somewhat misleading, as the case may be considered, not one of retention but rather of too frequent passing of urine.

Treatment.—Give a saline purge such as Epsom salts, three or four teaspoonfuls, and let the patient sit in a bath of hot water. If not relieved very quickly, then pass a *clean* catheter into the bladder, and allow it to empty itself (*see* Catheters, p. 254). After the bladder has been emptied put the patient to bed and give a dose of opium or bromide of potassium to procure rest. When he desires to pass water again let him have another bath, and if this is not effectual, again withdraw the water through a catheter. Patient should be careful to ward off further attacks by avoiding chills, over-drinking, and other exciting causes. If there is inflammation of the bladder, copaiba or sandal-wood capsules should be used; if the urine is irritating, bicarbonate of soda must be given.

Suppression of Urine.

In this serious condition no urine is secreted by the kidneys, so that on passing a catheter the bladder will be found to be empty.

Causes.—Shock from injury, inflammation and blocking up of the kidneys. Suppression of the urine is a rather common complication of severe cases of blackwater fever.

Treatment.—Hot baths, hot poultices to the loins, free use of aperients, especially Epsom salts and other saline purges. Bicarbonate of soda in full doses. Keep the skin acting freely by means of sweet nitre, or Warburg's tincture, or five-grain doses of antipyrine. Injections of hot water into the lower bowel.

Cystitis, or Inflammation of the Bladder.

Causes.—Injury or the result of operations, extension of inflammations such as gonorrhœa, retention and decomposition of urine; debilitated or gouty persons are especially liable to this affection.

Symptoms.—Intense pain in the lower part of the belly, and in the crutch, continual desire to pass water, with frequent passage of small quantities. The urine is scanty, high-coloured, foul-smelling, and occasionally blood-stained, and there may be some fever.

Treatment.—Hot baths, leeches, or fomentations to the crutch, and a sedative, such as opium (preferably given by the bowel), will be required. If the disease continues the bladder should be washed out through a catheter with weak boric acid solution, five grains to the ounce, or chinosol (1 in 2000), twice a day. Urotropin, ten grains, and copaiba or sandal-wood oil in ten-drop doses.

The diet should be restricted to milk.

Syphilis.

Syphilis, or the Pox, is an infectious venereal disease, nearly always communicated by direct contagion. The course of the disease is marked by a primary sore, the chancre; early constitutional (secondary) symptoms, and late constitutional (tertiary) symptoms.

In primary syphilis the disease is limited to the part or organ originally infected, and the glands connected with that spot. After an incubation period of from three to six weeks a small painless pimple appears at the seat of infection; it breaks down, and forms a small ulcer from which oozes a little watery fluid. The base of the ulcer and the skin surrounding it are hard like gristle. The nearest glands, usually those of the groin, enlarge and occasionally become tender. Unless badly neglected, the original sore gradually heals and the glands resume their normal size. Secondary symptoms now make their appearance. These are fairly definite, and comprises (*a*) A skin rash, consisting of numerous irregularly shaped copper-coloured spots, spread over the face, upper part of the chest, the loins and the back of the arms. They do not itch. (*b*) Moist lumps and warts form in the crutch, around the scrotum (purse) and the outlet of the bowel. (*c*) Ulcerated sore throat. Large deep ulcers form on each tonsil, having ragged undermined edges. (*d*) Iritis or inflammation of the eye may also occur. These symptoms, even if untreated, tend to heal, but always leave more or less marked traces behind.

The discharge from either primary or secondary sores is infectious and may convey the disease, so that great care needs to be taken in handling such sores.

After an interval of apparent health, lasting perhaps only a few months, but often for a year or two, the tertiary symptoms or "reminders" make their appearance. These take the form of localised swellings, which soon break down, forming deep ulcers, and if untreated, produce extensive destruction of the part involved, with much deformity.

Treatment.—As soon as the disease is recognised, the treatment must be commenced.

Local treatment.—Keep the sore perfectly clean by washing it with an antiseptic solution such as chinosol (1 in 1000). Between the washings, dress it with a piece of lint soaked in "black wash," or dust it with iodoform powder and cover it with a piece of lint smeared with boric ointment.

For the sore throat, use an antiseptic gargle (*see* Ulceration of Throat, p. 195).

General treatment.—The patient must be put on a course of mercury at once. Calomel, one grain twice a day, or grey powder, one grain three times a day, must be administered, and continued until skilled advice can be obtained. The effect of the mercury must be carefully watched, and if the patient complains of soreness of the gums, a coppery taste in the mouth and excessive flow of saliva, the dose must be reduced or the administration of the drug stopped until these symptoms have disappeared. If the calomel or grey powder causes looseness of the bowels, five grains of Dover's powder may be added to each dose.

In some cases, the addition of three grains of the iodide of potash to each grain of calomel does good from the very first.

For the later symptoms, continue the mercurial treatment, and give at least five grains of the iodide of potassium three times a day.

Gonorrhœa.

Gonorrhœa, or clap, is an acute inflammation of the urethra or pipe, attended with a discharge of more or less matter. It is nearly always due to direct contagion.

Symptoms.—At first there is some itching about the end of the pipe, which is followed by a yellowish-white discharge. This lasts from three to five days. Then great pain is noticed on passing water, and the discharge becomes thick and yellowish-green in colour, with redness and swelling about the lips of the opening of the pipe. After a time the pain on making water disappears, and the discharge becomes thin and watery, a condition known as "gleet."

Treatment.—Forbid alcohol in any form. Give large quantities of liquid—water, weak tea, or milk—to thoroughly flush the system. Light diet and as complete rest as possible. Keep bowels well open with saline and other purges. Give sandal-wood oil or copaiba, twenty drops three times a day, and urotropin, ten grains twice a day.

If there is much pain in the acute stage, a mixture containing fifteen grains of bicarbonate of soda, and five drops of chlorodyne or laudanum, in an ounce of water, may be given twice a day.

When the acute symptoms have subsided, the pipe must be syringed out with a very weak solution of permanganate of potash; later on, a lotion containing four grains of sulphate of zinc to one ounce of water may be used as an injection.

If the glands in the groin become tender and inflamed, they should be painted with tincture of iodine. If, in spite of this, the pain and swelling increase, they should be poulticed frequently, and treated as ordinary abscesses.

Piles.

Piles are very common in the tropics, and are often due to want of exercise, chronic constipation, dysentery, too free use of alcohol, over-eating, and excessive smoking. No one who suffers from piles should become a traveller till skilled advice has been obtained.

Internal Piles, though not usually painful, are by their frequent bleeding a cause of anæmia and debility; they lie inside the orifice of the bowel, but sometimes they come down on straining, and are then nipped by the muscle surrounding the opening, and may swell up, become very painful, and bleed profusely.

Treatment.—Keep the bowels freely but gently opened by taking cascara regularly; if the piles come down they should be returned, and an ointment of galls and opium or an injection of hazeline (one tablespoonful mixed with seven of water) used. Tannin, five grains to the ounce, or sulphate of iron, three to five grains to the ounce, may be used instead of hazeline. Hazeline suppositories are often of great use for internal piles, but ordinary suppositories do not keep well in very hot countries; if they are taken to the tropics they should therefore be specially made and packed. If the piles bleed profusely or cause great pain, an operation will be necessary.

External piles do not bleed, but from time to time they become inflamed and swollen, causing great agony.

Treatment.—The bowels should be kept well opened; the sufferer should lie with his hips raised, and hot fomentations should be frequently applied, and the piles should be well greased. Glycerine of belladonna, smeared on a pad of lint, is a valuable application.

Some sedative, such as Dover's powder, may be necessary to procure rest and sleep.

Burns and Scalds.

Where an extensive burn or scald has occurred, the clothing of the injured part should be removed by cutting, so as to cause as little irritation as possible. If the burn is only slight, the surface may be covered over with lint smeared with zinc or boric ointment, or oil. If there is much blistering, or the surface is charred, the skin should be cleaned up as well as possible with boric acid lotion, and hot fomentations of the same applied for twenty-four hours. After this, the burn may be dressed twice a day with boric ointment spread on lint. Great cleanliness is an important factor in the successful treatment of burns. In a severe burn, stimulants must be given, and the patient put to bed with hot-water bottles, and active treatment of the burn should be left till the patient has somewhat recovered from the shock.

When there is great pain, chlorodyne or laudanum in full doses will be required.

Blisters on the Feet.

These are generally caused by ill-fitting or badly finished boots, creases in the socks, or moisture of the feet. Great care should be used in the selection of boots and socks. Knitted socks are best, and they must be frequently changed. Rubbing the inside of the socks with soap is a valuable preventive.

When blisters have formed, they should be pricked—to let out the fluid—and adhesive plaster applied, so as to effectually protect the part. When rest can be obtained, zinc or boric ointment should be applied.

If the feet are naturally tender, and prone to form blisters, soak them for some time in tepid salt and water, or alum and water, before putting on the socks in the morning. Dust the feet, after they have been well washed and dried, with starch powder, or smear them over with zinc ointment; a pad of oiled lint, evenly strapped on, will often save a tender place from becoming raw.

TREATMENT OF WOUNDS AND INJURIES.

In treating wounds, whether large or small, the great essential is absolute cleanliness, not simple cleanliness in the ordinary acceptance of the word, but absolute surgical cleanliness.

In the first place, it is as well to lay down the axiom that all inflammation and other complications of wounds, as suppuration and blood-poisoning, are due to germs, and germs alone. All cleansing operations and antiseptic processes are instituted with the sole idea of preventing the entrance of germs into the wound, either from the hands of the operator, the dressings, instruments, or the skin of the patient.

Antiseptics are chemical substances which have the power of killing germs outright, or so checking their growth that the cells in the blood can easily cope with them. The antiseptics that are mentioned in these notes are carbolic acid, chinosol, permanganate of potash, and boric acid.

Given the antiseptic properties of the substance, it seems a simple thing, by its application, to prevent germs either entering a wound or multiplying there, but in actual practice the germ-free cleanliness of a wound is a difficult thing to procure. The germs are everywhere—on the instruments which cause the wounds, on the skin of the patient, in the sweat glands of the skin, at the roots of the hairs, and on the hands of the operator, especially under the nails. Again, a great obstacle to the free action of antiseptics is grease or oil of all kinds. A fine coating of grease enveloping a germ offers great resistance to the action of antiseptics, and the object of the process of cleansing a wound described below is, first, to remove all the grease, and then to apply the antiseptic to destroy the germs. With care, a clean-cut wound should heal kindly, without the formation of matter.

To clean and dress a Wound.

First cover the wound itself with a small piece of cotton-wool or lint, wrung out in an antiseptic solution; then thoroughly scrub the surrounding skin with warm water and soap (soft soap for preference), using a nail brush which has been previously boiled.

Shave the skin in every case, whether thickly covered with hair or not. This shaving is most important, as it removes all fine grease-coated hairs which might favour the growth of germs, and also scrapes off the surface dirt and dead scales of skin.

After shaving, again scrub with soap and water, and wash the skin with an antiseptic lotion, such as carbolic (1 in 60), or chinosol (1 in 1000).

Having cleansed the skin, now clean up the wound itself. Thoroughly wash the wound with an antiseptic lotion, either rubbed in with lint or wool, or injected with a *glass* syringe.

All instruments, ligatures and needles should be boiled for at least five minutes in water (to which a little washing-soda may be added with advantage), and should then be dropped into an antiseptic solution. When boiling is not possible, they should be soaked in carbolic lotion, 1 in 30, for ten minutes, and then transferred to 1 in 60.

The hands of the operator must be scrubbed for some minutes with soap and warm water, and then in an antiseptic solution, particular attention being paid to the nails, which should have been cut quite short. Cleanliness of the operator's hands is the first essential of successful treatment of wounds.

The edges of small, clean-cut wounds may be brought together with adhesive plaster, but larger wounds will need antiseptic silk or gut ligatures to keep the edges applied evenly.

Large ragged wounds, with much bruising; wounds containing dirt, sand, etc., or deep wounds such as those caused by bullets or spears, should not be completely closed. They should be cleansed, as above, dusted with iodoform powder, and a strip of antiseptic gauze arranged so as to reach from the surface of the skin to the deepest part of the wound. This, which serves as a drain for the discharges, should be renewed every day until healing is fairly established. If necessary, a few stitches may be inserted in the more superficial parts of the wound to bring the edges of the skin into contact.

The wound should be dressed with strips of antiseptic gauze, slightly moistened with whichever antiseptic solution is used, and over this should be placed a pad of antiseptic wool, which should be kept firmly and evenly in place by a well-applied bandage.

A wounded limb should be kept at rest, as far as possible, either by sand-bags or a splint.

(In the absence of other antiseptics, the wound, after thorough washing, may be dressed with carbolic oil or Friar's balsam, or boric or iodoform ointment applied on lint.)

Bleeding or Hæmorrhage,

General oozing from small vessels may be stopped by means of a pad made of a piece of antiseptic wool or gauze, firmly bandaged over the bleeding area.

When a large vessel, whether an artery or a vein, is cut across, profuse bleeding will take place, and immediate steps must be taken to stop this whilst suitable instruments are being obtained. (In bleeding from an artery, the blood spurts out in quick jerking jets ; if coming from a vein, the blood flows in a steady continuous stream.) Pressure should therefore be applied, by means of the thumb or thumbs, or a tourniquet,* in the course of the vessel, either above or below the injury—nearer to the body than the wound if the bleeding is from an artery (Fig. 3), and beyond the wound if the bleeding is from a vein.

It will be found that the bleeding can be controlled more effectually, and with greater ease, if the vessel is compressed against a neighbouring bone.

No more pressure should be exerted than is just sufficient to stop the flow of blood.

Whilst pressure is being applied, the wound should be cleansed with some antiseptic lotion, and a wedge-shaped pad of antiseptic gauze applied and firmly held in position by a bandage.

Bleeding from the hand or forearm can generally be immediately arrested by forcibly bending up the forearm at the elbow-joint.

---

* A tourniquet is a special instrument devised for the purpose of applying pressure to the main vessels of the limbs ; in its absence, one may be improvised by rolling a handkerchief into a narrow band, and inserting a stone or a cork between its folds to serve as a pad ; the pad is adjusted over the spot where it is proposed to compress the vessel, and the ends of the handkerchief knotted loosely round the limb. A stick is now slipped between the knot and the limb, and twisted round until sufficient pressure is exerted to arrest the bleeding.

If these measures effectually control the bleeding, the pressure should be kept up for an hour or two, after which time it may be cautiously relaxed. If, after the removal of the pressure, the hæmorrhage seems to have ceased (as judged by the pad which has been kept in position not trickling with blood), apply a large pad of wool *over* the original dressing, and bandage this firmly. Dress the wound very carefully on the third day.

If, on the other hand, in spite of the treatment, the bleeding continues, the pressure must be re-applied, and the cut ends of the bleeding vessel looked for in the wound itself, and either twisted or tied with a silk ligature.

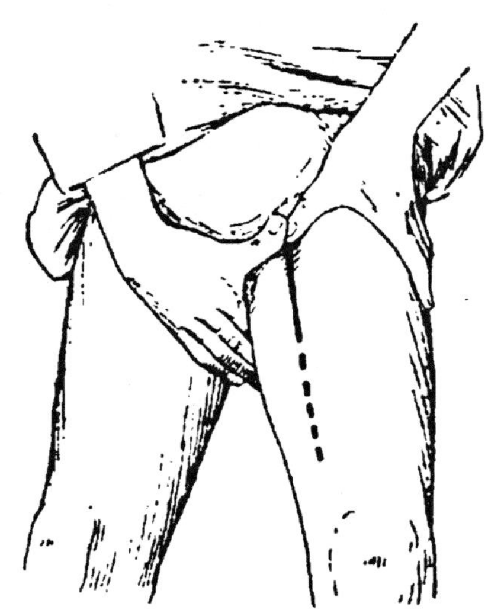

FIG. 3.—METHOD OF COMPRESSING THE MAIN ARTERY OF THE THIGH.

To twist a vessel, seize the bleeding point between the blades of the artery forceps, then, by twisting the instrument round three or four times, the vessel will become blocked or closed, and the forceps may be taken off in the course of ten minutes. The other method is to tie the vessel with a silk ligature (previously boiled), making a reef or sailor's knot, to cut short the ends of the ligature, remove the forceps, and let the vessel fall back into the wound. In any case cleanse the wound, and dress in the manner previously described.

The traveller who has taken the trouble to learn something about the treatment of wounds and severe hæmorrhage will, before his journeys are ended, probably have opportunities of saving the life of a comrade by his skill. Nothing can be more distressing than to see a man die from a hæmorrhage which anyone who has studied the subject a little would be able to control in nine cases out of ten without much difficulty. I therefore earnestly urge all travellers to gain as practical a knowledge of this subject as is possible, before leaving a civilised country.

After-Treatment of Wounds.

A clean-cut wound, which has been completely closed and properly dressed, need not be dressed again until the fourth day; and the stitches need not be removed for seven days.

On the other hand, a ragged or dirty wound, or one in which it has been necessary to insert a piece of gauze for the purpose of drainage, should be syringed out and dressed daily from the first.

The important index to the state of a wound is the patient's temperature. If, on the third day after the wound was first dressed, the temperature is normal, and subsequently remains so, the wound is probably healing well; but if the patient's temperature is up to or beyond 100° F., and the part is painful, there is probably some inflammatory change going on in the wound. The dressings should be taken off, and the wound examined. If the surrounding skin is red and shiny, and the limb swollen and tender, some of the stitches should be taken out, and the wound well syringed with an antiseptic lotion, This, with a good purge, such as calomel or Epsom salts, will probably remove all signs of inflammation; but still the wound must be dressed daily. If, however, the temperature still remains high, 100° to 101° F., or more, and the patient is restless and light-headed, all the stitches should be taken out, the wound opened up, and hot fomentations, prepared with an antiseptic lotion, applied every four hours till the temperature comes down. Tonics of quinine and iron should be given.

Erysipelas.

Erysipelas is a diffusely-spreading inflammation of the skin, caused by the presence and activity of germs, which enter through a wounded surface. It is most frequently due to want of cleanliness in the treatment of wounds. Bright's disease and gout predispose to this complaint.

Symptoms.—The disease begins with a vivid red blush, usually starting from a wound, and which has a great tendency to spread or to move from one part of the limb to another. The edge of the blush is sharply defined, and slightly raised above the surrounding skin; there is stiffness and heat in the part, with fever 103°–104° F., not varying much; rapid pulse, headache, loss of appetite, furred tongue and constipation.

Treatment.—Isolate the patient and administer a good purgative. Give twenty drops of tincture of steel, with five grains of quinine in two ounces of water. Light diet. Stimulants will be necessary. Local treatment: flour, starch, or zinc oxide may be used to dust over the part, or zinc ointment applied on lint. The healthy skin just beyond the advancing edge may be painted with tincture of iodine.

Cellulitis.—Occasionally the infection of the skin by the germs spreads to the tissue beneath, and is then called cellulitis, or phlegmonous erysipelas. In this condition there is more swelling than in simple erysipelas, and the skin is very boggy or doughy, and retains the imprint of a finger pressed upon it. The red blush is not so vivid, being darker and more purple, and there is no well-defined edge. If left untreated, the skin will break down and die, causing extensive destruction of the part.

Treatment.—General treatment the same as for erysipelas. Locally, several incisions, each at least one inch in length, should be made into the swollen tissue where it is most boggy, and right through the skin, care being taken to avoid the larger blood-vessels; then hot fomentations of boric acid (five grains to the ounce) or other antiseptic should be applied, every two or three hours, till the swelling has subsided. The important point in the treatment of swollen and inflamed parts following wounds, when accompanied by continued or rising high temperature, is to relieve tension by making free incisions. Some knowledge of the anatomy of the parts is essential before using the knife. In any case, incisions in the arm or leg should be made parallel to and not across the limb.

Blood-Poisoning.

If a poisoned wound is left without thorough local treatment, the poisons and germs contained in the tissues are taken up by the blood, and cause *blood-poisoning*. The symptoms of this condition are high temperature, delirium, headache, loss of appetite, vomiting, and occasionally bronchitis and pneumonia. Severe forms of blood-poisoning may lead to the formation of abscesses all over the body, with fatal result.

Treatment.—The general treatment is the same as that described for erysipelas. The local treatment is to thoroughly open up and disinfect the poisoned part.

Bruises.

A bruise should be treated by bathing with cold water, or the frequent application of wet cloths. The addition of Goulard water, gin, methylated spirits, eau de cologne, or vinegar, to the water, is beneficial.

Sprains.

The affected joint should be raised on pillows, and treated in the manner advised for bruises, but if seen immediately after the injury, firm strapping with adhesive plaster and bandaging of the part is often equally serviceable. If inflammation develops, warm fomentations will be found soothing, leeching may be necessary, and free purgation always has a good effect.

The troublesome stiffness which often remains is relieved by friction and kneading with the hand. To regain the use of the joint, it should be gently moved each day; this movement is less painful if it is performed with the joint in hot water.

Poisoned Wounds from Snakes, Animals, and Arrows, etc.

In cases of poisonous snake-bite, the marks of two fangs will usually be found.

Treatment.—Tie a piece of tape, bandage, or cord a couple of inches above the wound, *i.e.*, between the wound and the body; tie another piece still nearer the body, say three inches from the first. Cut across the wound or wounds to encourage free bleeding.

(If a medical man is present, he may think fit to inject anti-venomous serum as an antidote to snake poison.)

Dissolve as much permanganate of potash as possible in about a teaspoonful of water, stirring well to hasten its solution. Inject about five drops of this underneath the skin, on either side of the cut, by means of a hypodermic syringe. Some of the solution may be injected into the wound itself, or even a crystal of the drug may be pressed into the cut, or part of a crushed soloid of the drug.

It is best to keep a small bottle of the strong solution of permanganate always ready.

Give spirits strong, that is, one tablespoonful to one of water; at least four such doses in the first hour. Ammonia may also be given.

If the patient is heavy and stupid, give two tablets of strychnine (one-hundredth of a grain in each) in half a wineglassful of water, by the mouth; or dissolve two in twenty drops of water, and inject well beneath the skin into the muscles of the back. If there is no improvement within an hour, give two more tablets; and if necessary, one or two more in another hour.

After tying up a limb for a poisonous bite, there will be great pain if the ligatures have been applied tightly; the parts will swell, become very dusky, and if the ligatures are left on too long, the blood supply will be cut off, and the parts will die. Therefore, when the ligatures begin to cause much pain, loosen the one next the injured part for ten seconds, then tie again at or near the same place, and loosen the other one for a similar period, and then re-tie; repeat this about every fifteen minutes. In the course of two hours, both tapes may be taken off. The object of tying up is to prevent much of the poison getting into the system at once.

The first thing is to tie up tightly; next cut freely, and suck or squeeze out as much blood as possible, then treat with drugs.

(It is usually safe to suck a poisoned wound unless there are any abrasions or cracks in the mouth, tongue, gums, or lips, taking care to spit out the poison at once; but the mouth should be washed out immediately afterwards with a light purple solution of permanganate of potash.)

If the wounds are on the face, neck, or other spot which cannot be tied up, then it is best to cut out the part at once and wipe the wound well with the strong solution of permanganate.

In the treatment of snake-bite, pure carbolic acid, ordinary caustic, a red-hot wire, or even a burning stick may be applied to the wounds when permanganate of potash cannot be obtained. In some cases, where no other treatment is available, it is advisable to explode a pinch of gunpowder over the place of injury, or even to blow the parts away with one's gun.

If a finger or toe is bitten by a snake which is certainly poisonous, and neither drugs nor fire are at hand, it would be best to amputate at once.

Wounds inflicted by poisoned arrows or other weapons, mad dogs, jackals, etc., should be treated in a similar manner to those caused by poisonous snakes.

Drowning.

Death from drowning usually occurs in from two to three minutes after submersion, although people have been revived after a period of five or six minutes under water.

In treating cases of apparent death from drowning, the points to be aimed at are:—

First and immediately, the restoration of the breathing.

Secondly, and *after breathing is restored*, the promotion of warmth and circulation,

1. *To restore the breathing.*

Roll the patient on to his face for a few seconds, placing one of his arms under the forehead; wipe away all weeds, mud, etc., from the mouth. (In this position water will more easily escape from the mouth, whilst at the same time the tongue will fall forward, and leave the entrance to the windpipe clear.)

Turn the patient on his back, on a flat surface, with the head a little higher than the feet.

Place a small hard pillow (or a rolled-up coat) *under the shoulder-blades.*

Draw the patient's tongue forward, and keep it projecting beyond the lips.

Remove all tight clothing from about the patient's neck and chest; also braces, belt, etc.

Kneel at the patient's head, grasp his arms just above the elbows, draw them gently and steadily upwards above his head, and keep them stretched in that position for two seconds (Fig. 4). (By this means air is drawn into the lungs.)

Reverse the movement, and press the patient's arms gently but firmly against the sides of the chest, keeping them in this position for two seconds (Fig. 5). (By this means air is pressed out of the lungs.)

Repeat these movements alternately and regularly, about fifteen times a minute, until natural breathing takes place, or as long as there is any hope of saving the patient. It may be necessary to continue the movements for as long as an hour.

Fig. 4.

While these movements are being carried out the wet clothing may be removed, the body gently dried, and the patient wrapped up in dry blankets.

Fig. 5.

2. *To promote warmth and circulation.*

Rub the limbs and body vigorously with dry towels and flannels, always rubbing from the extremities towards the heart.

Apply hot-water bottles, hot flannels, or hot bricks to the feet, armpits, the pit of the stomach, and between the thighs.

Immediately the power of swallowing returns, administer hot coffee, spirits and warm water, etc.

Put the patient to bed between hot blankets as soon as possible.

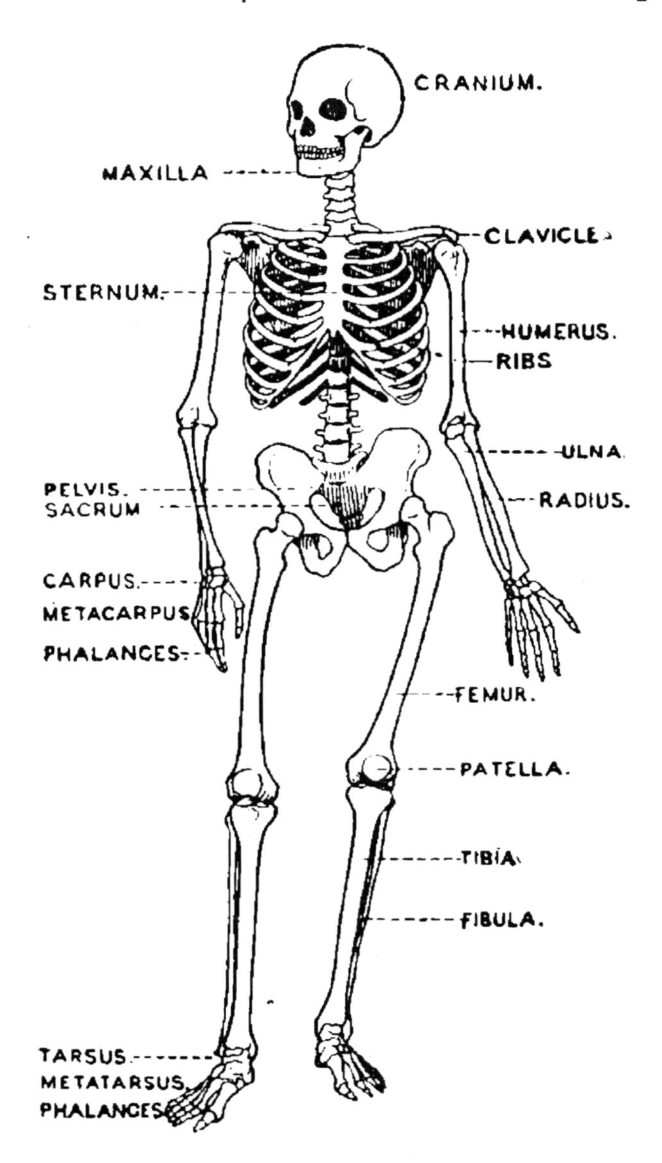

FIG. 6.—DIAGRAM OF THE HUMAN SKELETON, GIVING THE NAMES AND POSITIONS OF THE CHIEF BONES.

Fractures.

A *simple fracture* is one in which, though the bone or bones are broken, the protecting skin is *not* broken.

A *compound fracture* is one in which the skin, etc., is broken or cut across, so that the fracture is more or less exposed to the air. The end of either fragment of the broken bone may protrude through the skin.

A *comminuted fracture* is one in which the bone is broken into several pieces; such a fracture may be either simple or compound.

In a simple fracture great gentleness should be used in handling the parts, so as not to convert it into a compound fracture; therefore, do not undress the part; rather cut away the clothing.

A fracture which is compound is usually serious, for dirt and germs are liable to be carried into the wound and cause great mischief; in gunshot wounds, dirt or pieces of clothing may be carried into the wound.

The signs of a fracture are, firstly, the patient's own feelings, *e.g.*, the pain which is caused on handling the part, sudden loss of power, and the sensation of grating. Secondly, on examination, most if not all of the following signs will be observed: inability of patient to move the part below the injury, swelling, unnatural movement below the site of fracture, alteration in appearance of the limb.

In fractures of the upper or lower limb there is usually shortening, the spasmodic action of the muscles causing the broken ends to ride over each other, and the greater the obliquity of the line of fracture the greater will be the shortening and alteration in appearance.

A sensation of grating is usually conveyed to the operator when he attempts to move the parts; sometimes this grating can be heard, as well as felt. Comparison of the injured with the sound limb is of the greatest importance in detecting fractures.

Treatment of Simple Fractures.

Directly a fracture of a limb is made out, a splint or splints of some kind should be applied to keep the parts fairly in position, and to prevent a broken end from being pushed through the skin. Cloths and bandages may be applied firmly round the injured part, and then extemporised splints, such as boards, straight sticks, umbrellas, or bayonets, should be applied and kept in position till the patient is in bed.

Splints suitable to the injury should be now made and well covered with wool, lint, or cloth, special care being taken to pad them well where they are likely to press upon bony prominences, such as the inner or outer ankle. The limb must next be straightened, and any deformity caused by overriding of the fragments must be remedied by steady pulling upon the parts above and below the fracture, in opposite directions, and the parts brought into good position by manipulating the bones at the seat of fracture. The prepared splints, extending well below and above the fracture, are then fitted to the limb with cotton-wool or lint, and secured by bandages.

If there is great swelling and tenderness of a limb, then it is advisable not to apply splints at once. Sand pillows should be made, or stockings nearly filled with sand and their mouths tied, and applied one on either side of the limb, to keep it absolutely still; the painful swelling may be reduced by applying ice or evaporating lotions. When the swelling is somewhat reduced, splints may be applied after proper manipulation.

Whilst the bandages should be firmly applied so as to keep the splints in position, they must not be bound too tightly, otherwise swelling and ulceration may be caused. If there is much pain and swelling after a fracture has been set, it will be necessary to loosen the bandages.

Union of the fractured bones is generally completed in about six weeks.

Collar-bones.—Fracture of the collar-bone should be treated by placing a large wedge-shaped pad (about six inches long, by three in thickness at the upper end) in the armpit, and securing it with tapes tied over the opposite shoulder. The elbow should then be brought forward, and raised and well supported by a broad triangular bandage or handkerchief, used as a sling, and with the ends tied over the opposite shoulder. A flannel or other bandage should then be wound round the chest, so as to secure the arm from accidental movements.

Ribs.—Fracture of the ribs may be treated by wrapping a flannel bandage round the chest pretty tightly, so as to limit the movements of breathing, which are very painful. The flannel should be secured by stitching, and the upper turns should be fixed by broad tapes passed over the shoulders and firmly stitched. Firm strapping of the side with adhesive plaster is still better. For this purpose about six strips of plaster one to two inches wide and eighteen inches long should be applied evenly round the side of the chest; each piece should be overlaid by the next piece above it for about half an inch. To secure rest for the affected side of the chest, the strapping should not only cover the broken bone, but should extend to about three inches above and below it, and should reach well beyond the middle line both in front and behind.

Upper Arm.—Fracture of the upper arm may be treated by the application of several narrow splints reaching from the armpit to the elbow, well padded, and supported in position by a bandage carried from the fingers to the armpit. Care must be taken that the splints on the inner side do not chafe the folds of the armpit. The hand and wrist should then be supported in a sling, but the elbow must be allowed to hang free.

Forearm.—Fractures of the forearm must be treated by two splints, each wider than the limb. The injured limb is allowed to hang down by the side, palm forwards. One splint reaching from the elbow to the finger-tips is applied to the back of the limb; the other is placed on the front, and only reaches from the bend of the elbow to the level of the ball of the thumb. The splints are secured temporarily by a couple of slip knots. Now bend the arm to a right angle, thumb uppermost, and bandage securely from the tips of the fingers up to the elbow.

Thigh.—Fractures of the thigh are serious; they require the patient to be kept in bed till union has been effected, and they are more likely to lead to shortening and permanent lameness if not very carefully treated, and the assistance of a skilled surgeon is urgently needed. A long splint is applied to the outside of the limb, reaching from the armpit to beyond the foot, and secured above by a bandage passing round the body, whilst the foot and leg are firmly bandaged to it below.

Leg.—Fractures of the leg should be treated by applying a splint on each side, long enough to reach from the knee to a little below the sole of the foot. They should be carefully applied with bandages, keeping the great toe in a line with the inner border of the knee-cap. When the accident occurs in the open air, the injured limb should be tied to the sound one, till the patient is brought to a place of security, the toes being prevented from pointing inwards.

Lower Jaw.—Manipulate the parts into their normal position and mould a splint of gutta percha, or other material, as accurately as possible to the lower jaw. If a tooth is loose and prevents the two jaws meeting properly it should be taken out.

Apply the splint and keep it in position by a bandage, which should be split at the chin so as to encircle the point of the jaw; the ends of the bandage are also split as far forward as the angle of the jaw; two ends are tied behind the neck and two over the top of the head, as in the diagram, and these tied ends should be united by a bandage or tapes to keep them in position (Fig. 8).

In the absence of suitable material for making a splint, this bandage alone will have to suffice.

Fig. 7.

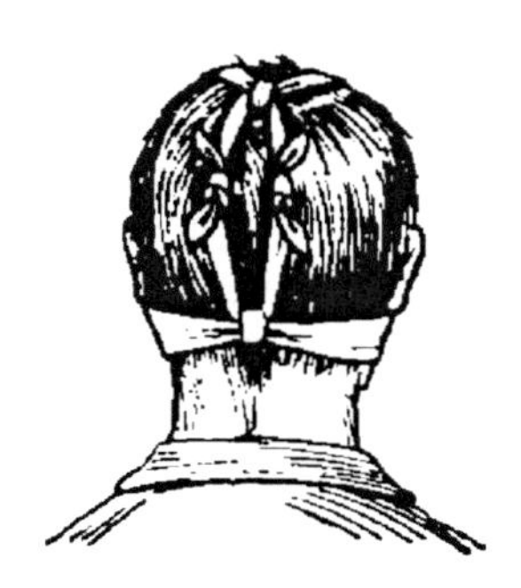

Fig. 8.

The patient must be fed on slops through a tube passed behind the teeth, or through a passage resulting from the loss of a tooth.

Compound Fractures.—Clean up thoroughly as described in the treatment of wounds. Wash out with an antiseptic solution, dust with iodoform, apply an antiseptic pad. Set the limb in such a manner that the wound can be dressed daily without interfering with the splints.

If the bone is protruding through the wound, it must, if possible, be got back into position. If this cannot be done it may be necessary to saw off the end of the bone to enable the wound to be closed.

If the wound is large, deep, or dirty, it should be stuffed with antiseptic gauze so that free drainage may be allowed, and the gauze should be changed each day.

After-treatment of Fractures.—Gentle movements may be cautiously begun, in the joints above and below the injury, in about three weeks. While these are being carried out, the limb must be firmly supported to avoid interfering with the healing processes going on between the ends of the bones.

Dislocations.

Dislocations nearly always require skilled aid to reduce them.

Shoulder-joint.—Patient cannot raise the arm to his head or perform any other shoulder movements freely. The shoulder is flattened, the elbow sticks out, and the limb is usually lengthened.

Treatment.—The patient should lie down. The operator removes his own boot on the same side as that of the patient's dislocation, inserts his heel into the patient's armpit and draws the arm steadily down, at the same time pressing the heel in an outward direction; the dislocated bone should slip back into its proper position. Put a pad in the armpit and bandage the arm to the side for a week; support arm in a sling for another week or two.

Elbow-joint.—Dislocation of the bones of the forearm backwards at the elbow-joint is fairly common.

Treatment.—This dislocation can usually be reduced by placing the knee in front of the patient's elbow, and making firm traction on the forearm—which is at the same time bent a little around the operator's knee. The patient can be kept sitting in a chair while this is done, and the operator can get his knee into the required position by placing his foot on the side of the chair.

Hip-joint.—Is the most common form of dislocation. The limb is shortened, bent at the knee and twisted inwards, the great toe of the injured limb resting on the instep of foot of the opposite limb. The outer side of the hip is swollen and distorted.

Treatment.—Grasp the ankle with one hand and the knee with the other. Lift up the leg and bend it at the hip, then carry it as a whole away from the other limb as far as possible, rotate the toes and foot firmly outwards, and in that position bring the limb back parallel to the sound one.

After-treatment of Dislocations.—After dislocations, gentle movement of the joints should be begun in two weeks, so as to prevent them becoming fixed.

POISONS.

It is impossible, in the small space available, to give anything like a detailed scheme of the treatment of the various poisons, but a few hints on the essential points must not be omitted.

I have therefore arranged a list of the commoner poisons, together with the special treatment appropriate to each. Following this are a few hints more especially adapted to the requirements of travellers.

Poison should be suspected (*a*) when severe and alarming symptoms of illness suddenly attack a person apparently in good health; (*b*) if the symptoms closely follow the taking of food, drink, or medicines; (*c*) if several people are attacked after having partaken of the same food or drink.

The points to be attended to in the treatment of cases of poisoning are:—

a. Remove as much of the poison as possible from the stomach.

To do this, it is usual to give an emetic, such as mustard and water, or zinc sulphate, thirty grains; or to pass the soft rubber tube down the gullet, and wash out the stomach with water.

b. Counteract the effect of the poison by an antidote.

c. Keep up the patient's strength till the poison is removed from the system.

d. Relieve pain.

For greater convenience, poisons may be divided into three classe according to their mode of action.

1. Corrosives, which, when swallowed, attack, burn, and destroy the lining of the mouth, throat, and stomach, *e.g.*, aqua fortis (nitric acid), oil of vitrol (sulphuric acid), spirits of salts (hydrochloric acid), oxalic and carbolic acids, etc.

Note.—In cases of poisoning by corrosives, the stomach tube, if used at all, must be introduced with extreme care. Antidotes must be chiefly relied upon.

Sulphuric acid. Hydrochloric acid. Nitric acid.—Antidotes: half an ounce of chalk, whiting, or even plaster from the walls or ceiling, thoroughly mixed with half a pint of milk. Soap-suds, gruel, or linseed tea may then be given freely. Later on, chlorodyne or laudanum may be needed to relieve the pain.

Carbolic acid.—The stomach tube may be used, but with extreme caution. Antidotes: sulphate of magnesia (Epsom salts), half an ounce, or sulphate of soda, one ounce, added to three ounces of water. Stimulants must be given, and the patient wrapped in warm blankets, and hot-water bottles applied. Emetics are useless.

Oxalic acid.—The stomach tube may be used, but with care. Emetics should be administered, but plain water must not be given. Antidote: chalk, half an ounce in five ounces of water. Stimulants must be administered, by the bowel if necessary.

2. Irritants, which do not at once cause destruction of the lining of mouth, throat and stomach, but subsequently irritate and inflame them, *e.g.*, arsenic, phosphorus, lead, bad meat, etc.

Arsenic.—Use the stomach tube or give emetics. Antidote: solution of dialysed iron, one ounce, repeated every two hours for twelve hours. Administer milk, linseed tea, barley water or white of egg and water freely. Stimulants must be given. Hot bottles and blankets.

Phosphorus.—Stomach tube. Emetics, especially sulphate of copper, five grains in half a pint of water. Antidote: crude French oil of turpentine, twenty drops. Epsom salts, half an ounce in three ounces of water, as a purgative. Avoid fats and greasy substances.

Corrosive sublimate.—(Perchloride of mercury.) If possible, do not use the stomach tube. Give emetics and milk freely. Antidotes: white of egg and water, flour and water, or arrowroot and water. Stimulants and hot bottles and blankets.

Croton oil.—Stomach tube or emetic. Stimulants. Opium enema, laudanum, twenty-five drops by mouth. Hot bottles and blankets. Poultices to belly.

Paraffin oil.—Stomach tube or emetic. Stimulants. Hot bottles and blankets. Friction.

3. Nerve Poisons, which act on the nervous system and cause symptoms such as headache, drowsiness, or convulsions, *e.g.*, alcohol, strychnine, opium, etc.

Alcohol.—Stomach tube and emetics. Apply cold water to the head. One pint of coffee, hot and strong, should be given by the mouth or lower bowel. Solution of ammonia may be held cautiously under the nose.

Opium.—(Chlorodyne, laudanum, morphia.) Stomach tube or emetic.

Strong hot coffee, one pint by mouth or bowel. Sal volatile or carbonate of ammonia, well diluted, may be given freely. Antidote: permanganate of potash eight grains in an ounce of water, give one teaspoonful every five minutes; or tincture of belladonna, thirty drops. Pinching, flapping with towels. Make the patient walk about. Artificial respiration if the patient is insensible.

Strychnine.—(Nux vomica.) Stomach tube or emetic, put patient under chloroform or ether; or administer bromide of potassium, two drachms with chloral thirty grains, in two ounces of water. Artificial respiration.

Belladonna.—(Deadly nightshade.) Stomach tube or emetic. Castor oil, one ounce, as a purge. Hot coffee. Warm and cold douche alternately. Hypodermic injection of morphia. Hot bottles and blankets. Artificial respiration. Friction.

Travellers may be poisoned in the following ways:—

1. By poisonous bites, arrow wounds, etc.
2. By eating food poisoned by natives.
3. By eating poisonous plants, etc.
4. By eating bad meat or fish.

The treatment of *snake-bites and arrow wounds* has already been dealt with.

Poisoned food and poisonous plants.—Food is occasionally poisoned by natives, either by adding strophanthus leaves or other poisons. The eating of such articles may be followed by severe pain, colic, vomiting, diarrhœa or other symptoms. The general line of treatment should consist in giving emetics promptly, washing out the stomach by means of the stomach tube, and administering large quantities of water, milk, or gruel. Distressing symptoms should be treated as they arise; thus, if convulsions occur, large doses of chloral or potassium bromide should be given; if stupor or drowsiness comes on, the patient should be forced to take hot coffee, stimulants, etc., and made to walk about. If collapse supervenes, stimulants should be freely given, the patient wrapped in blankets, and hot bottles applied to the feet.

Poisoning by bad meat, fish, etc.—The symptoms of poisoning by bad meat, etc., are severe pains in the belly, cramps, and shivering fits, followed by vomiting and diarrhœa coming on soon after its consumption, occasionally within an hour or two. The vomiting and diarrhœa are usually alarming and very severe, causing great collapse, and unless urgent remedies are applied, death may occur. The bodies of patients may be covered all over with irritating blotches, red rashes or with nettle rash.

Treatment.—Give an emetic at once. Administer a sharp purge to clear out the bowels. Use stimulants freely. Apply hot-water bottles and blankets. Give opium or chlorodyne, fifteen drops every four hours until the pain is relieved. Milk and gruel or arrowroot.

MEDICINES, MEDICAL APPLIANCES, ETC.

Weights and Measures.

Solids.

			Symbol	Gr.
	A Grain		Symbol	Gr.
20 Grains	1 Scruple	weights now rarely used	„	℈
60 Grains	1 Drachm		„	ʒ
437·5 Grains	1 Ounce		„	℥
16 Ounces	1 Pound		„	℔

Note.—An ounce weighs not 8 but rather under 7½ drachms.

1 Gramme	about 15½ grains.
1 Kilogramme	„ 2 lbs. 3 ozs.

Liquids.

1 Minim	About 1 drop	Symbol	♏
60 Minims	One fluid drachm.	„	ʒ
8 Fluid drachms (480 Minims)	One fluid ounce	„	℥
20 Fluid ounces	One pint	„	O

A teaspoonful	About one fluid drachm.
A dessertspoonful	About two fluid drachms.
A tablespoonful	About half a fluid ounce.
A wine-glassful	About two fluid ounces.
A tumblerful	About half a pint.
A litre	About a pint and three quarters.

All bottles containing drugs should be doubly labelled, and the labels should be varnished, otherwise they will probably come off in a damp climate.

Drugs in a liquid state are inconvenient for travellers. They are bulky, and require very careful packing and handling.

As far as possible, therefore, I have selected drugs which are put up by Messrs. Burroughs, Wellcome & Co. in the form of tabloids and soloids.

Particulars concerning a special case of drugs supplied by the above firm, designed for the tropics, will be found on p. 41 of "Hints on Outfit." The Station Chest of Messrs. Howard Lloyd & Co., of Leicester, contains a good assortment of crude drugs.

Tabloids are often taken in the solid form, but they will act more quickly and certainly if dissolved or suspended in about an ounce of water.

Some preparations, chiefly those used for preparing lotions, etc., are called soloids, *e.g.*, the soloids of chinosol. These are really tabloids.

In regard to drugs, I have kept well within the maximum dose, so that there may be no trouble on that account. For instance, in certain cases a medical man would give sixty drops of laudanum for a dose, but I advise travellers to rarely give more than sixty drops in twenty-four hours. The urgency of the case must be the guide as to the quantity of the drug to be given. The smallest dose which is effective is the best.

A supply of antiseptic gauze wool and lint is necessary.

All antiseptic gauzes and other dressings should be very carefully protected from the air in order that their properties may be preserved. They should be wrapped in oiled paper or india-rubber tissue, and kept in a tin box.

A supply of bandages will also be needed. Each bandage should be two inches in width by six feet in length.

Dressing case.—The traveller should provide himself with a surgical dressing case. A suitable and inexpensive one is made by Messrs. Down Bros., St. Thomas's St., London Bridge; it contains 1 pair of scissors, 1 pair of Spencer Wells' artery forceps, 1 probe, 1 scoop and grooved director, 1 knife with two blades, 1 small saw with a detachable handle, and a packet containing silk, wire, needles, and pins.

All active poisons mentioned in the following pages are marked with an asterisk (*).

Aloin Co. tabloids.—One taken three times a day after meals, in chronic constipation, gradually reduced as bowels become regular.

Alum.—Dose, five to ten grains. Is occasionally used as an internal astringent in the treatment of diarrhœa.

A solution containing five grains to the ounce of water may be used as a mouth wash for bleeding or inflamed gums, as a gargle for relaxed and sore throats, or as a lotion for inflamed eyes.

Ten to twenty ounces of a solution containing five grains to the ounce is sometimes used as an enema to check the diarrhœa in chronic dysentery.

The "soloid" of alum weighs ten grains.

Ammonia.—Three preparations of ammonia, viz., Sal volatile, Carbonate of ammonia, and Liquor ammoniæ, are extremely valuable as stimulants, and in this respect they are to be preferred to alcohol. All three are similar in their action, and are valuable on account of their stimulating effect upon the heart in cases of fainting, or collapse caused by snake-bite, bullet-wound, or other injury; they relieve spasm, and promote sweating in feverish states. They also cause free expectoration, and are therefore useful in bronchitis.

The chief objection to these drugs is that their strength is rapidly lost, unless they are kept in well-stoppered bottles, and in the dark.

All preparations of ammonia should be administered in about one ounce of water.

Sal volatile, or aromatic spirit of ammonia.—Dose, twenty to sixty drops for a single administration, or up to thirty drops if repeated frequently. This is the most convenient and pleasant preparation of ammonia, but is bulky. As a local application, it is of great value in relieving the pain caused by the bites of certain insects, *e.g.*, mosquitoes and fleas.

Carbonate of ammonia.—Dose, three to ten grains.

Liquor ammoniæ, or solution of ammonia.—Dose, five to twenty drops.

(*Note.*—This is not the strong solution, which is always labelled "Liq. Ammon. Fort.," and which is three times stronger than liquor ammoniæ.)

As a local application, the liquor ammoniæ, diluted with an equal quantity of water, may be injected into the wounds caused by snake-bites.

Ammonium chloride.—Dose, five to twenty grains. This drug is chiefly used in congestion of the liver, especially where acute inflammation or abscess is threatened; at times it is useful in neuralgia and chronic bronchitis. It is also used for inflamed and relaxed sore throat, in the form of a gargle, five to ten grains to the ounce of water.

Antipyrine.—Dose, five to ten grains. (Uses as Phenacetin.)

* *Araroba, or Goa powder.*—Is not for internal use. One part of the powder mixed with one of acetic acid and fourteen of lard makes what is known as araroba ointment.

Araroba ointment is useful in many skin diseases, such as psoriasis; it easily destroys the parasites in ringworm, dhobie itch, or kraw kraw. It is a powerful drug, and should be used only in small amounts, for if applied too freely it will cause some inflammation and pain.

* *Aromatic chalk with opium tabloids.*—Ten grains three times a day in severe cases of diarrhœa.

* *Arsenic.*—Dose, one-sixtieth to one-fifteenth of a grain. Is a powerful and poisonous drug, and should be taken with caution. It is useful in chronic malarial poisoning, and in anæmia and debility, especially that following an attack of fever.

It is best combined with iron in tabloid form—iron and arsenic tabloid—two of which may be taken twice a day, after food, for a week. This is a good general tonic.

Bismuth carbonate.—Dose, five to twenty grains. Useful in dyspepsia, pain in the stomach, vomiting and diarrhœa. The drug is heavy, and therefore gum or starch water should be used to suspend it.

It is often combined with opium and bicarbonate of soda to check vomiting and diarrhœa. One ounce of gum water, to which has been added two drops of carbolic acid, five grains each of bicarbonate of soda and carbonate of bismuth, and ten drops of laudanum, is useful in cases of dysentery, profuse diarrhœa and cholera.

Bismuth may also be used as a dusting powder.

Boric acid, or boracic acid.—Chiefly used in solution, for its mild antiseptic properties, as a gargle, injection, eye lotion, or mouth wash. On the same account, it is used for washing wounds, or syringing out ulcers, or sore and inflamed ears. It is practically non-poisonous.

Lotions should contain from five to ten grains to the ounce. Boric ointment is a useful dressing for sores, burns, or wounds; it may be made by adding one part of powdered boric acid to nine of vaseline or fat.

One part of finely-powdered boric acid added to six parts of starch makes a good dusting powder for moist eczema, bedsores, chafes, and perspiring feet.

Boric lint (lint impregnated with boric acid) is useful as a dressing for wounds, ulcers, and abrasions.

Boric wool is absorbent cotton-wool impregnated with boric acid, and is the most generally useful antiseptic wool.

Bromides.—The bromides generally used are those of potassium, or ammonium. Dose in either case, five to twenty-five grains. They are sedatives, and are used in convulsions, epilepsy, and delirium; they relieve headache, especially if taken with sal volatile, and may be combined with chloral (*see* Chloral). The bromide of ammonium is less depressing than the bromide of potassium.

Calcium sulphide, i.e. sulphurated lime.—Dose, a quarter to one grain three times a day. Useful for those who are suffering from boils or carbuncles.

* *Calomel, or subchloride of mercury.*—Dose, two to five grains. Calomel may be used in almost all cases requiring an aperient; its action will be assisted by giving fruit salt, or citrate of magnesia, four hours later. In cases of insensibility or delirium it is specially useful, for the powder may be placed on the back of the tongue, and it will be swallowed unconsciously.

Certain cases of dysentery are cured by calomel.

Four grains of calomel may usually be given with advantage, at the beginning of a malarial attack.

In syphilis, one grain may be given two or three times a day; but its action should be watched, and if it causes salivation, tenderness of the gums, and a coppery taste in the mouth, it should be discontinued until these symptoms disappear.

Camphor.—Dose, two to five grains. Is a stimulant, and an anti-spasmodic; it helps to calm the nervous system. It is used in cholera and in fevers, and is very valuable for colds in the head.

Camphor is slightly soluble in water, and this solution makes a cooling lotion, which is useful for inflamed or painful parts.

Spirit of camphor consists of one part of camphor in ten of alcohol; the dose is ten to twenty drops.

Camphorated oil consists of one ounce of camphor in four ounces of olive oil, and makes a stimulating liniment for stiff and painful parts. A stronger liniment may be made by adding to this an ounce of spirit, and half to one ounce of solution of ammonia or oil of turpentine.

* *Carbolic acid, phenic acid, or phenol.*—Dose, one or two grains may be given internally, well diluted, for acute diarrhœa, dysentery, and cholera. Is mostly used as an external application. Carbolic acid acts as a caustic, and may be applied on the end of a match to poisoned wounds, *e.g.*, snake-bite, arrow wounds, and parts stung by venomous insects. A drop on a piece of cotton-wool applied to a hollow tooth quickly relieves pain.

It is much used as an antiseptic lotion for wounds and foul sores, and for cleansing hands and instruments before operation.

One part in thirty of water makes a strong solution for cleansing instruments and disinfecting dysenteric and other stools.

One part in sixty may be used for sponging or syringing sores and wounds.

One part added to nineteen of olive, or other oil, makes carbolic oil, which is a handy dressing for all kinds of wounds.

One part to nineteen parts of vaseline, or fresh lard, makes carbolic ointment.

Cascara sagrada, extract of.—Dose, two to six grains. Is a valuable aperient, especially in chronic constipation. Cascara tabloids, two grains each, are convenient, and one or two should be taken at night.

Cathartic Co. Tabloids.—A useful purge in cases of fever or liver derangement. Two or three for a dose when required.

Castor oil.—Dose, one to three tablespoonfuls. It is a good aperient, but is not very convenient for the traveller, as it takes up much space, which could be better used for aperients which are not so bulky, *e.g.*, calomel or cascara.

Note.—The seeds of the castor oil plant should not be eaten, as they are poisonous.

Chinosol is a drug which has many of the advantages of carbolic acid without its poisonous or caustic properties. It is so generally useful that, at any rate for explorers, it will largely replace carbolic acid. It is antiseptic and disinfectant in its action, can be used as a mouth wash or gargle for sore gums and ulcerated throats; it makes a good lotion for washing the hands, instruments, or wounds, and may be syringed into fresh wounds or into chronic ulcers.

It is put up in the "soloid" form by Burroughs, Wellcome & Co., each soloid weighing eight and three-quarter grains, and the strength of the solutions here given are calculated strictly from this weight.

One soloid dissolved in a pint of water makes a solution of 1 in 1000 (equal to about 1 in 40, carbolic lotion), which is the most useful for general purposes, such as an antiseptic wash for the hands or for disinfecting surgical instruments.

A soloid in two pints of water makes a solution of 1 in 2000. This may be used for washing fresh wounds, burns, and suppurating surfaces, or as a gargle for sore throat.

As a healing dusting powder, one part of chinosol may be mixed with ten of boric acid, and used in the same manner as iodoform powder.

To disinfect typhoid or dysenteric stools, dissolve four soloids in one pint of water and add the mixture to the vessel containing the motions.

* *Chloral.*—Dose, five to twenty grains, Relieves restlessness and delirium, and produces sleep. Larger doses than twenty grains should not be given. In severe convulsions, due to certain poisons, *e.g.*, strychnine, twenty grains or more of bromide of potassium may be added to the full dose of chloral, and given either by the mouth or bowel.

* *Chloroform.*—Dose, one to five drops. Two to four drops can be given with advantage with almost any drug, as a flavouring agent, and on account of its antispasmodic action.

From two to four drops put on a piece of loaf sugar and sucked will often stop sea-sickness or other vomiting.

Spirits of Chloroform, or Chloric Ether, consisting of one part of chloroform in twenty parts of alcohol, is a convenient form in which to administer the drug. It is useful in sea-sickness, cramps and colic, and should be given in doses of twenty drops added to a teaspoonful of water every quarter of an hour, until six such doses have been given.

* *Cocaine.*—A one per cent. solution may be used to diminish pain in the eye caused by injury or a piece of grit lodging under the lids. Two drops should be applied to the eyeball, and in about two minutes the application may be repeated; in another minute the sensibility of the part will be diminished, and a careful examination may be made for dust or other foreign bodies, which should of course be cautiously removed.

Creosote.—Dose, one to two drops; is best given in a capsule. In many cases of indigestion it gives great relief if administered immediately after food. Creosote applied to a hollow tooth, upon a piece of cotton-wool, will usually relieve the pain.

* *Croton Oil.*—Dose, half to one or two drops, mixed with bread crumb or in a pill.

Is a powerful purgative not usually given to Europeans. Native carriers like it.

Copaiba, Balsam of.—Dose, five to twenty drops three times a day. Is best taken in capsules. It increases the expectoration, and is therefore useful in bronchitis; is a stimulating disinfectant to the urine, bladder and pipe, and is used in gonorrhœa and inflammation of the bladder. It may produce an irritating rash on the skin if taken in too large quantities.

Emetics.—The following are the usual emetics:—

Common salt, two tablespoonfuls in half a pint of water.
Mustard, one tablespoonful " "
Ipecacuanha, thirty grains " "
Zinc sulphate, thirty grains " "

In the absence of any of the above, give copious draughts of tepid water, followed by tickling the back of the throat with the finger or a feather.

Enema.—*Nutrient enema:*

Two eggs,
Half teaspoonful of salt,
A wine-glassful of milk.

Beat up the eggs, then add the salt and milk, and mix well. Inject slowly through the long tube well up into the bowel.

Nutrient enema:

Benger's beef jelly, two tablespoonfuls,
Milk, about three ounces.

Warm-water enema. To relieve uneasiness of lower bowel, as in dysentery, piles, etc., half a pint or more of warm water, by the short tube.

Soothing enema. Laudanum, thirty drops; warm water about two ounces. A little thin starch water or gum may be added. Inject through the short tube, and repeat the injection in three hours if patient is not easier.

Quinine enema. Quinine, twenty grains or more; warm water, about two ounces. If the patient does not retain the injection, give a second one within an hour. Select a soluble preparation of quinine, such as the bisulphate or the hydrobromide.

Ipecacuanha and opium enema. Ipecacuanha, thirty grains; laudanum, twenty drops; starch or gum water, about two ounces. Especially useful in dysentery, when the stomach will not retain ipecacuanha.

Aperient enema. Wash hands in thirty ounces of warm water, using plenty of soap to make a good solution; or use plain warm water or weak gruel; the addition of one or two tablespoonfuls of oil is of advantage. Inject slowly into the bowel with short tube; if it does no cause uneasiness, inject the whole amount. The injection should be retained as long as possible.

Epsom salts, or sulphate of magnesia. Dose, two teaspoonfuls or more. Is a saline purgative, giving a copious watery motion, rapid in its action, and useful in fevers or congestion of the liver. It also increases the flow of urine.

Recently has been much used in dysentery (p. 181).

Friar's Balsam.—(Tinct. Benzoin Co.) Dose, thirty drops or more.

Internally is useful as an expectorant in bronchitis.

Externally, lint soaked with it is an efficient dressing for wounds and sores. When first applied it causes a good deal of smarting, but this soon passes off.

Gall and opium ointment is a useful application for piles.

Gauze, antiseptic. The best dressing for wounds is some form of antiseptic gauze. Cyanide gauze, which is impregnated with cyanide of mercury, is the most generally used.

Ginger, essence of.—Dose, five to twenty drops. It is useful in cases of cramp, colic, and indigestion, especially if combined with five or ten grains of bicarbonate of soda and two or three drops of chloroform.

If there is much pain, ten drops or more of chlorodyne or laudanum may be added.

* *Goa powder.*—*See* Araroba.

* *Goulard water.*—For external use only. Is astringent, and may be used as a lotion for inflamed joints. Is also useful as an injection in gonorrhœa, and gleet, as a soothing lotion in herpes, shingles, eczema, discharge from the ear, and itching and chafes of all kinds. *It should not be used as an eye lotion.*

Goulard water is made by adding one part of Liquor Plumbi Subacetatis Fortis, or Goulard *Extract*, and one of alcohol to seventy-eight of water.

One of the soloids of the subacetate of lead added to a quarter of a pint of water (distilled when convenient) makes a solution similar to Goulard water.

Gum acacia is used for forming a mucilage with which to suspend such drugs as carbonate of bismuth. One part of gum to three of water should be made and strained before use. Starch may be used for the same purpose.

Gum water may be very freely given in cases of irritable bladder.

Hazeline.—Dose, thirty to one hundred and twenty drops. It is a powerful astringent, very useful in spitting of blood. May be applied freely to bleeding parts, such as the nose, gums, piles, or small wounds. One tablespoonful of hazeline to five of water may be injected into the bowel for bleeding piles.

Ichthyol.—An ointment containing twenty per cent. of ichthyol is very soothing in most skin diseases, such as eczema and psoriasis. The ointment is of service in reducing the pain and swelling of mosquito bites.

Iodide of potassium.—Dose, five to ten grains in the later stages of syphilis.

* *Iodine, tincture of.*—Dose, two drops in a teaspoonful of water; given every hour, is most valuable for obstinate vomiting.

Used externally, as it acts as a counter-irritant, and should not be painted on too freely as it may cause blistering. Does good if painted over swollen joints or enlarged glands, but should not be applied if the skin is broken. Tincture of iodine may be painted on the chest or over the liver or spleen if there is pain.

Is very useful in certain diseases of the skin, such as ringworm, kraw kraw, and dhobie itch.

Iodoform powder is a valuable antiseptic, and is used for all kinds of sores, or wounds, in which it rapidly promotes healing. One part of iodoform to eight parts of vaseline makes a good healing ointment.

Ipecacuanha.—Dose, expectorant, half to two grains.
,, emetic, fifteen to thirty grains.
,, for dysentery, twenty to forty grains or more.

In small doses, *e.g.*, a quarter to half a grain, is a stomachic and may check vomiting.

Is much used in dysentery (p. 178).

By causing free expectoration, it is very useful in bronchitis, and is then usually combined with ammonia.

Is contained in Dover's powder, and in this form is useful in coughs and colds, as it helps to cause sweating.

Iron.—Iron is much used on account of its tonic and blood-forming properties. It is especially useful in anæmia following fevers and other exhausting diseases, and it also gives tone to the nervous system.

Most preparations of iron are astringent in their action, some such as the sulphate and perchloride being more so than others; these are, therefore, useful in diarrhœa and in bleeding from the gums and bowel. Iron by its astringent action may cause constipation, and it may be necessary to add a small amount of an aperient, such as Epsom salts, to each dose.

The following preparations are those generally used:—

Tincture or perchloride of iron, or tincture of steel.—Dose, five to fifteen drops in at least an ounce of water. Good blood tonic. Is very astringent, and is therefore useful in internal bleeding and looseness of the bowels. It is of very great value in erysipelas.

Sulphate of iron.—Dose, one to three grains. Is a powerful astringent and blood tonic, and often does good in checking the looseness of the bowels which remains after an attack of dysentery.

Iron pill or Blaud's pill.—Dose, five to fifteen grains. Is one of the best tonic preparations of iron, as it is not very astringent. It is put up in tabloid form.

Solution of dialysed iron.—Dose, ten to thirty drops. Is a good blood tonic, and, unlike most preparations, is not astringent.

Parrish's food.—Dose, thirty to a hundred and twenty drops in water. Is a good tonic preparation of iron.

Iron and arsenic compound tabloids.—A useful tonic in convalescence from malaria, one taken three times a day after meals.

Mercurial ointment forms a useful application in cases of itch, ringworm and other skin diseases. As it is rather strong, it should be diluted with two parts of vaseline.

White precipitate, or ammoniated mercurial ointment. — Uses as mercurial ointment.

* *Opium.*—Is a valuable drug, but it is also a poison, and great care should be observed in using it. Opium is narcotic and sedative in its action; it relieves pain of all kinds. Useful in diarrhœa, dysentery, cramp and colic. Promotes perspiration and checks vomiting. Sometimes it will cut short a cold or an attack of malaria.

No preparation containing opium should be given to children.

The chief preparations containing opium are:—

Chlorodyne.—The ordinary dose is ten to fifteen drops, but if there is great pain, then even thirty or forty drops may be given, but a second dose should not be administered for three or four hours; if two full doses have been given, do not give a third within at least twelve hours of the second dose, and do not give it at all if the patient is drowsy.

It is best not to give more than forty to sixty drops in twenty-four hours, unless there are very special reasons for giving a larger quantity. Is the most suitable of all the preparations containing opium. Is to be preferred to laudanum, as it is more palatable and more readily stops vomiting.

* *Laudanum, or tincture of opium.*—The doses and uses are precisely similar to those of chlorodyne. Laudanum has been put up in the form of a tabloid, and is very convenient for transport.

Dover's powder, or compound ipecacuanha powder.—Dose, five to fifteen grains. It contains opium and a small amount of ipecacuanha. Ten grains of Dover's powder are equal to about fifteen drops of chlorodyne or laudanum. The tabloid is an extremely handy preparation. Dover's powder is especially useful in coughs and colds, the ipecacuanha it contains assisting the action of the opium. If a patient who is chilled is put to bed at once, kept warm with blankets and hot-water bottle, and is given ten or fifteen grains with a hot drink, he will perspire, and probably the cold will be cut short.

Peppermint, spirits of.—Dose, five to twenty drops. Is a valuable drug in cases of flatulence and dyspepsia, especially if combined with bicarbonate of soda; it also masks the taste of unpleasant medicines. For children, one drop of the spirit and one grain of bicarbonate of soda may be added to one teaspoonful of water, and this dose may be frequently repeated.

Peptonic tabloids, a useful digestive tabloid, taken with food when digestion is weak, or may be used to predigest it.

* *Perchloride of mercury soloids*, for making lotions for disinfecting wounds, etc.

Phenacetin.—Dose, three to eight grains. Phenacetin is used for headaches, or for reducing the temperature in malaria and other febrile diseases, and to cause sweating. As it is very insoluble, it is best taken in the form of a powder, or in alcohol and hot water.

Potassium, bromide of. *See* Bromides.

**Potassium permanganate* in solid form is a mild caustic, and is the active principle of Condy's fluid, which contains about eight grains of the drug in one ounce.

It is disinfectant, deodorant, and antiseptic.

One part of this drug is soluble in about twenty parts of water.

One grain gives a purple colour to a gallon of water. Impure water turns the purple rapidly to a brown colour, therefore the permanganate is a rough test for the presence of organic matter in water.

A pale purple solution is useful as a gargle or mouth wash, also as an injection in gonorrhœa; wounds may be cleansed with a similar solution.

It is especially useful in poisoned wounds, such as snake-bites, and should in these cases be injected hypodermically (p. 251).

Is an antidote to poisoning by opium (p. 232).

Purgatives may be divided into the following classes:—

1. Laxatives.
2. Simple purgatives.
3. Drastic purgatives or cathartics.
4. Saline purges.

1. Laxatives slightly stimulate the movements of the bowel, but cause practically no increase of the intestinal secretion. Examples of this class are fruit, wholemeal bread, small doses of castor oil, figs, prunes, etc. These may be given regularly in slight cases of chronic constipation.

2. Simple purges are more powerful than laxatives, and in addition to stimulating the movements of the bowel increase the secretion. The result is a soft, easy motion. Examples: rhubarb, senna, cascara sagrada.

3. Drastic purgatives or cathartics. The drugs cause a profuse flow of the intestinal secretions, and, occasionally, of the bile, with greatly increased bowel movements. They cause a copious watery evacuation, accompanied by a good deal of griping. Examples: calomel, colocynth, aloes, podophyllin, croton oil, jalap and oil of turpentine; of these, podophyllin and aloes cause an increase in the flow of bile, with increased intestinal movements, so they may be said to act directly upon the liver. The result is a profuse, watery, bile-stained motion.

Calomel does not act directly upon the liver, but stimulates the upper part of the bowel, so that its contents are hurried along before the bile can be reabsorbed, and a loose, watery, bile-stained motion ensues.

The most useful purgative of this class is the pill containing colocynth, calomel, and hyoscamus, of which five to ten grains should be given when there is a furred tongue, constipation, heaviness or weight over the liver, and dyspepsia and loss of appetite.

4. Saline purgatives. These cause a free secretion of the intestinal juices, and a copious motion, proportionate to the size of the dose, is the result. The principal saline purges are Epsom salts (*i.e.*, sulphate of magnesia), seidlitz powder, the various forms of fruit salts, and the aperient mineral waters, such as Rubinat, Hunyadi Janos, etc.

A saline purgative may be given in the morning, to assist the action of an aperient pill administered the previous night.

Saline purgatives are useful, too, in many cases of habitual constipation, and, if necessary, a small dose may be given every morning in a tumbler of warm water.

Quinine.—Quinine is chiefly used for malarial fever, and the urgency of the case must be the guide as to the amount to be administered. It is seldom necessary to give more than ten grains three times a day; at times enormous doses, such as thirty grains three times a day, have been given with advantage, but such doses are rarely necessary, and in the absence of a medical man should never be given.

The administration of quinine sometimes causes headache, deafness, delirium, and partial or even complete blindness. In such cases the dose should be reduced or the drug withheld until these symptoms have disappeared.

It has been stated that quinine causes blackwater fever; this, I believe, is absolutely untrue. I have seen cases of blackwater fever, apparently resulting from the neglect of malarious attacks, which date the commencement of their recovery from the first administration of quinine.

Quinine, like all other drugs, acts most quickly if given in solution; the nauseous taste can be somewhat disguised by adding chloroform water or essence of ginger or lime juice to the solution. Many men take the drug in a little beer or spirit for the same purpose.

Quinine pills are, as a rule, very insoluble, therefore they are not suitable for explorers. Quinine in the form of a tabloid, or in a capsule, or wrapped in tissue paper, acts well.

Sometimes, owing to frequent vomiting or other cause, the quinine administered by the mouth cannot be retained. It must then be given by means of an enema or by injection beneath the skin.

If the traveller decides to give a hypodermic injection of quinine, he should only give five grains at a time. The dose should be added to about a teaspoonful of water which has been boiled and allowed to cool. The amount of water is of little importance so long as enough is used to dissolve the drug, and keep it in solution when the water is cool enough to be injected.

Injections are best made deeply under the skin or into the muscles of the flanks, sides, or back, and not into the limbs.

The preparations of quinine.—Many preparations of quinine are made; they are all popularly known as quinine, their uses are similar, and the doses are as follows:—

Dose, one to five grains, as a tonic.
„ five grains as a preventative of malaria.
„ five to twenty grains, in fevers.

Sulphate of quinine, formerly called quinine disulphate.—This is the preparation most commonly used; but it is very insoluble in water, and is, in point of fact, the least suitable.

Acid sulphate of quinine, sometimes called bisulphate, soluble or neutral sulphate of quinine.

The acid sulphate is much more soluble than the sulphate, and is a more suitable preparation for administration by the mouth or for injection into the bowel.

Hydrochloride of quinine.—Is a very soluble preparation, and contains a large percentage of quinine.

Quinine hydrobromide.—Is more soluble than the sulphate, and less liable to cause headache and other quinine symptoms.

Acid quinine hydrobromide.—Very soluble and non-irritating, therefore is one of the best forms for hypodermic use. Solubility, one in six of water.

Lactate of quinine.—Is a soluble preparation, and is suitable for hypodermic injection. I have never found it cause irritation when injected well under the skin. Solubility, one in ten of water.

Acid quinine hydrochloride, or quinine dihydrochloride, sometimes called quinine bihydrochloride. Is a very soluble preparation, and is suitable for hypodermic injection.

It is soluble in less than its own weight of water.

For administration by the mouth or bowel, the traveller is advised to take either the hydrochloride or the hydrobromide of quinine.

For hypodermic administration, I formerly used the lactate of quinine; recently I have only used the acid quinine hydrochloride.

Rhubarb, either in form of compound rhubarb pill, one of the most useful of ordinary aperients, two to four for a dose.

Compound Rhubarb Powder, otherwise known as Gregory's Powder.

Sandalwood oil.—Uses and doses as for balsam of copaiba.

Soda, bicarbonate of.—Dose, five to ten grains or more. Five to ten grains dissolved in water may be given twice a day for acidity, flatulence, heartburn. Useful in dysentery to allay irritation of stomach; for this purpose five or ten grains may be given an hour before each dose of ipecacuanha.

If the urine is acid and scalding, give twenty grains three times a day. Bicarbonate of soda is given in cases of blackwater fever when the urine is not sufficient in quantity; in such cases at least twenty grains should be given every six hours.

Soda mint tabloids, useful for relieving flatulence, one or two being taken after meals.

Soda salicylate.—Dose, ten to twenty grains. Relieves pain in rheumatic fever. Lowers temperature. Its action must be carefully watched, as too large doses may cause headache, buzzing in the ears, and even delirium.

Sulphur.—Precipitated sulphur is useful in diphtheria; an ointment containing one part of sulphur to nine of lard, grease, or vaseline, makes a useful application for itch and other skin diseases.

Tannin.—Tannin is a powerful astringent, and may be used internally or externally.

In five to ten grain doses it helps to stop diarrhœa, bleeding from stomach, bowel, piles, and bladder. If there is continued bleeding, increase the dose to twenty grains, and repeat it as often as necessary.

Voice tabloids, consisting of chlorate of potash, borax and cocaine are useful in cases of sore throat.

Warburg's tincture.—One tabloid is equal to thirty drops of the ordinary tincture.

To produce sweating, first open the bowels, then give four to eight of the tabloids with a little hot spirit and water. If necessary, repeat the dose in three hours, and in either case withhold if possible food or drink till after this period has elapsed, but if the patient is very thirsty give hot drinks of weak tea or spirit and water. Keep him covered up and put a hot-water bottle in the bed.

Urgency of the case is a guide as to how many tabloids ought to be given.

Zinc oxide alone or mixed with an equal quantity of boric acid or starch makes a useful drying dusting powder.

Zinc ointment contains three parts of zinc oxide to seventeen parts of lard or vaseline. This is a most useful application for burns, eczema, chafes and sores.

Sulphate of zinc.—As a prompt emetic, give twenty to thirty grains in half a pint of warm water. If patient does not vomit give a pint or more of warm water and tickle the back of the throat. A second dose of the drug may be given.

Two grains to an ounce f water makes a valuable astringent lotion for gonorrhœa, raw surfaces, ulcers, and inflamed eyes and ears. The strength of this solution may be gradually increased to six grains to the ounce of water.

TEMPERATURE TAKING, HYPODERMIC INJECTION, BATHS, CATHETERS, STOMACH TUBES, ENEMAS, AND POULTICES.

Pulse, Respiration and Temperature.

Normal pulse rate (adult) about 72 per minute.

Normal respiration rate (adult), 15 to 18 per minute.

Normal temperature, 98·4° F., but varying in health between 97·5° and 99°.

Temperature Taking.

The temperature of the body may be taken in the armpit, the mouth, or the lower part of the bowel.

In the Armpit.—Dry armpit well, put the bulb of the thermometer into it, and press the arm tightly against the body, so that the thermometer will be in close contact with the skin on either side for at least five minutes.

In the Mouth.—Wash the thermometer in an antiseptic solution, dry it and place the bulb well under the tongue and close the lips on the stem; at least three minutes should be allowed. This is a more reliable way of taking the temperature than in the armpit.

In the Bowel.—If the patient is very ill and light-headed, it is not advisable to place the thermometer in the mouth, and when a patient is having a cool bath to reduce his temperature a reading will be needed at least every quarter of an hour; in these cases it is best to pass the end of the thermometer, which has been previously greased, about two inches up the bowel, and to leave it there for at least two and a half minutes.

Do not trust in half-minute or minute thermometers, always allow the full time given above. Before taking a patient's temperature see that the mercury in the thermometer is shaken down at least as low as 97° F. After the temperature has been taken, note it down carefully, wash the thermometer, and shake the mercury down to 97° F. If the thermometer is washed in hot water, the expanding mercury may break it.

Hypodermic Injection.

The best place to inject is under the skin of the back or chest. The forearm is usually selected, but there are many objections to it. If the injection is made into the leg or arm, the point of the needle should be directed towards the body.

1. Surgical cleanliness must be observed. The hands of the operator and the skin of the patient at the spot chosen for injection must be properly cleaned with an antiseptic lotion, *e.g.*, chinosol (1 in 1000) or carbolic acid (1 in 60).

2. The hypodermic syringe and needle must be perfectly cleaned; they may be boiled, or an antiseptic lotion, or alcohol (brandy or whisky), syringed through them.

3. The solution to be injected should be made in a clean, *i.e.*, boiled, spoon, by adding the drug to boiled water.

4. Draw solution into the syringe, hold the syringe with the needle end pointing upwards, so as to allow any air to rise above the fluid, and push the piston up till all air has been driven out and the solution begins to come through the needle.

5. Pinch up the skin where the injection is to be made, push the needle well through and then under it, keeping the point slightly away from the skin until the needle is almost entirely covered. The point of the needle will now lie in the loose tissue—between the skin and the muscle—into which the fluid is to be injected.

6. Holding the needle firmly, slowly press the piston until the required amount has been injected; withdraw the needle slowly, keep the finger over the minute opening in the skin, so as to close it at once, and with another finger rub for a few seconds, where the fluid has been injected, in a direction away from the opening, to assist its absorption.

After giving an injection, wash out the syringe as before, dry needle well, and pass a piece of thin wire through it to keep it open. The needles are sent out with wire in them to keep the fine channel open; of course the wire must be removed before the needle is put on to the syringe.

As strong solutions of permanganate of potash attack the plunger of the piston, when this drug has been injected, the syringe should be well washed with water directly after its use.

Hypodermic tabloids.—Solutions for injection are best made from tabloids which are specially prepared for the purpose. A hypodermic tabloid should be dissolved in about ten drops of hot water, but as each contains a definite amount of the drug, it does not matter in exactly how much water the dose is dissolved. In injecting certain preparations of quinine, which are not easily soluble, so much water may be required that two or three separate injections are necessary; there is no objection to this, the important thing is to have the quinine really dissolved.

Usually only medical men should administer hypodermic injections. I advise others, if they do so, only to inject half or a third of the amount of the fluid in which a hypodermic tabloid has been dissolved. As a rule, it is not necessary to give a hypodermic injection, for the hypodermic tabloid will be absorbed, and act almost as quickly if placed under the tongue.

The traveller may be called upon to use one of the following hypodermic injections:—

Permanganate of potassium. *See* Snake-bite, p. 221.

Quinine. *See* Malaria, p. 167, and Quinine, p. 247.

Morphia. An injection of a quarter of a grain of morphia may be necessary when there is much pain. The soothing action of opium and chlorodyne is due to this substance.

Strychnine. *See* Snake-bite, p. 221.

Baths, etc.

Cool Bath.—Lift the patient gently in a blanket into a long bath containing water at about 90° F., add water as cold as can be obtained, keep the water circulating and running out; one boy must pour water on to the head all the time.

If a long bath is not at hand, put a blanket over a waterproof sheet, and let the patient lie on it; get six or eight boys to hold up the edges, and give patient his bath in that way.

A good plan is to dig a shallow trench in the sand and to spread a waterproof in it; it is less difficult then to keep the edges of waterproof up as the patient is lying in a groove. It is not necessary to make the bath really cold if the cool water can be kept running. I have hardly ever been able, in Africa, to get the water below 80° F.

Note the bowel temperature, and as soon as it falls to 101° F., remove the patient from the bath and put him between warm blankets, and feed with some hot soup. If there are any signs of faintness, ammonia, or equal parts of strong spirit and water, must be given. Weak spirit and water has little or no stimulating effect on the heart, therefore, in cases of fainting, shock, exhaustion, or collapse from any cause, give a little spirit, and give it strong.

Wet Pack.—Slip a waterproof under patient, wring a sheet out of cold water and pass it under him and wrap him in it, cover with three or four blankets, and tuck him in. At first he will feel cold and chilly, but soon the surface vessels will dilate, and he will begin to feel hot, and very likely will sweat; keep him in the wet pack with the clothes carefully tucked in round the neck, feet and sides for about half an hour. Then partially dry the patient, and put him into bed between warm blankets and with hot-water bottles, and give hot drinks to encourage perspiration.

Catheters.

At least six silk-webbed catheters should be taken as part of the equipment. The most suitable sizes are, 2, 3, 4, 6, 8, 10. They should be packed in a separate box, and should be kept dry with French chalk or any other dusting powder.

Before use the catheter should lie for at least a quarter of an hour in an antiseptic solution such as chinosol (1 in 1000), and the solution should be frequently syringed through it. The catheter should then be lubricated with eucalyptus, vaseline, or carbolic oil, or, better, with boric acid ointment.

Let the patient lie down, and thoroughly wash the genitals, especially the opening of the pipe, with an antiseptic solution, the operator having carefully cleaned his own hands. Then pass the catheter gently down the pipe and into the bladder. The silk-web catheters are so flexible that unless force is used they can do no harm; begin with a No. 8 size, and if this cannot be passed, try a No. 4.

After use, wash the catheter as before, dry with a soft handkerchief, and dust it with powder. On no account should the catheters be greased when they are being put away. Neglect of cleanly precautions in passing catheters may lead to inflammation of the bladder and even more serious mischief.

Stomach Tube.

The stomach tube consists of a piece of india-rubber tubing about three feet long, with a funnel at one end. The method of passing it is as follows:—The patient's jaws should be kept apart by a gag such as a piece of stick wrapped in lint placed between the back teeth on either side. The left forefinger should be passed into the patient's mouth as far as possible and the tongue drawn forward; the rubber tube, oiled or smeared with vaseline or glycerine, should then be passed by the side of the finger and down the throat for about twenty inches, taking care to use no great force. The funnel is then raised and slowly filled with water; when all the water has run down, the funnel is lowered and the liquid in the stomach is drawn out by the syphon action of the tube. The process should be repeated till the water returning from the stomach is clear.

If the patient is unconscious the tube should only be passed when he is lying with the feet raised above the level of the head, or water may be poured into the lungs and endanger the patient's life.

Enemas.

Enema Syringe.—This is fitted with long and short tube. To use the long tube, slip it over the short one, which will hold it firmly. After use hang it up to drain, dry it well, but do not oil it; carry it loose rather than coiled up, so as to avoid risk of the rubber kinking at the flexed portions of the tube.

To give an enema, the patient should be placed on his left side, and brought close to the edge of the bed, with his knees slightly drawn up towards the belly. The pipe or nozzle of the syringe should be well oiled or smeared with vaseline, and then carefully introduced into the outlet of the bowel, and passed gently upwards for about three inches, great care being taken not to exert any force.

The higher up into the bowel a nutrient or medicinal injection is passed, the more rapidly will it be absorbed; therefore the long tube is to be preferred, and the hips should be raised on a pillow, to assist the patient in retaining the injection.

A nutrient or medicinal enema is usually small in quantity, about two to four ounces, in order to prevent its being rejected by the bowel, and is administered by means of a small ball syringe provided with a long nozzle.

Before giving a medicinal or nutrient enema, it is best to wash the lower bowel with warm water, and always see that the mixture to be injected is warmed to "blood heat."

Poultices.

Linseed-meal poultice.—Mix four ounces of linseed meal into about half a pint of boiling water, constantly stirring until the mixture is smooth and even. A piece of tow, teased out to the required size, or a piece of linen or thin cloth, is placed upon a table, and the poultice turned out upon it; then spread evenly in a layer about three-quarters of an inch thick, leaving a margin of tow or linen about an inch wide all round. This margin should be folded over, and the poultice applied to the affected part—with the meal next the body.

An *Ice poultice* is made by mixing pounded ice and sawdust, and enclosing the mixture in a waterproof material such as a mackintosh or gutta-percha.

MOUNTAIN TRAVEL.—*By* DOUGLAS W. FRESHFIELD.

Revised by CLINTON T. DENT.

The Highlands of Central Asia, and the ranges of western North America, are among the fields likely next to attract explorers. If their exploration is to be thorough, travellers must take with them some knowledge of glacial phenomena. They must learn to know glaciers and moraines when they see them, to distinguish between ice and *névé*, permanent and temporary snowbeds. They must also be able to climb summits sufficiently high to command the recesses of the chain and the secrets of the snow world. In order to do this, they must be at the pains to acquire at least the rudiments of the mountain craft which has been brought to perfection by three generations of Alpine peasants. Without these qualifications, even surveyors will find themselves obliged to leave large and, to the physical geographer and geologist, singularly interesting tracts of country ill-mapped and imperfectly explored, and they will run the risk of bringing away very erroneous and incomplete impressions of the phenomena of great mountain chains. The practised mountaineer is free both from the fears and the rashness of the less experienced traveller, or the native of the Himalayas, the Andes, or the Caucasus. He is not likely to be deterred from visiting a remote valley because ice and snow, and possibly steep and rocky ridges (held impassable by the native hunters), intervene between him and it; on the other hand, he will not start on such an enterprise without every appliance that may enable him to conquer the difficulties of the way; he will not walk across a *névé* without a rope; he will not be frightened into retreat by the first crevasse, or stopped by a hard-frozen slope. If possessed of such mountaineering qualities, he will command the confidence of the natives on whose assistance the results of his journey largely depend, and he will be able to go in safety where others might not be able to go at all.

Ropes and ice-axes, procurable at Hill's, 4, Haymarket, are essential. It is still more essential that their proper use (up to the present time hardly known outside Europe) should be learnt. This may best be done in an Alpine tour, with an experienced glacier guide. Travellers without such experience had best keep to frequented passes, or below the snow-level. They will be most in danger when they perceive it least, and will imperil the lives of themselves and their companions. Icecraft, like seamanship, has to be learnt. A party of three is the smallest consistent with safety above the snow-line; and, whatever the number, the majority in any expedition of difficulty, should be experienced climbers. Such expeditions will best be made from a base where the heavier luggage and attendants are left.

The best scheme for mountain exploration is one which neither limits the traveller to a single district or valley, nor carries him straight on from point to point, but allows for various short expeditions from a succession of centres, at which he can leave his camp and heavy luggage.

The effect of rarefied air at great heights in reducing the powers of the human frame is a subject on which precise knowledge is still wanting. Probably no one has yet closely approached the limit at which the exertion of walking uphill becomes impossible to a person in normal health and accustomed to great elevations. It lies, therefore, considerably above 23,000 feet. On the other hand, mountaineers agree that their powers diminish perceptibly as they ascend above 12,000 feet. In De Saussure's generation both he and his guides were, at 15,000 feet, on Mont Blanc, unable to do more than advance a few yards at a time, while men of science now spend three days and nights on the summit of Mont Blanc, and modern climbers feel little or no inconvenience 2000 feet higher on the difficult peaks of the Caucasus, and can still climb and observe between 22,000 and 23,000 feet in the Karakoram or the Andes. Probably up to 18,000 feet the body acclimatizes itself to the upper air; and "training" is therefore one of the best preventives of mountain-sickness. The inhalation of oxygen has been advised as a palliative, but the remedy, if such it be, is not practically possible on the mountain side. The inconveniences felt on high ascents arise in some part from indigestion, and light but frequent meals (*e.g.* soup at starting, peptonised meat sandwiches and chocolate and cold tea during the climb) will be found very efficacious in avoiding bodily discomfort. A scientific investigation of the process by which the human frame adapts itself to high altitudes has recently been made by M. Vallot. (*See* Levasseur's 'Les Alpes,' Paris, 1889; Geographical Journal, January, 1893; Sir Martin Conway's work on the Karakoram, and Mr. Whymper's on the Andes.) Professor

Mosso, in his 'Life of Man on the High Alps' (English translation, Fisher Unwin, 1898), gives the results of an elaborate investigation into the subject, but many of his views have been seriously questioned. The subject is complex, involving both local and personal conditions, and demands further experiment and research; all dogmatic statements must at present be received with reserve.

Next to the rarity of the air, frostbite is the most formidable enemy of the climber who attempts great altitudes. Satisfactory foot-gear has not yet been devised. Some modification of Arctic expedients suitable for rock-climbing is wanted. The feet must not be compressed and the circulation impeded. Generally foreign mountaineers pay more attention than Englishmen to climbing-shoes and crampons. The ordinary hob-nail is good enough for most places where an explorer ought to go, but crampons undoubtedly enable their wearers to reach a point which would be unattainable to them by stepcutting, and much time is often saved by their use, if all the party are provided with them. It is extremely important that the crampons should fit the boot accurately. They can be obtained of the Albion Iron and Wirework Co., Red Lion Street, E.C. The straps should be of hempwebbing, not leather ¾ inch wide.

Mr. Whymper's tent is probably the best pattern to use for high climbing. It is much improved by the addition of a fly, which adds little to the weight. Alpine sleeping-bags, snow spectacles, felt-covered water-bottles, self-cooking souptins, chocolate, warm covering for hands and feet, strongly nailed and easy boots, cloth gaiters, are among the chief requisites for high exploration. A complete list of outfit is given in a 'Report on the Equipment of Mountaineers,' to be obtained from the Assistant Secretary of the Alpine Club (price 6*d.*). Various preventives of sunblistering have been advised. Professor Mosso states that soot is the most efficacious application. Cold cream, zinc ointment and the like, prevent any bad degree of sunburn. Take plenty of spare snow spectacles for use by porters in crossing snow passes. Field-glasses are much appreciated as presents by most mountain people, and spare ones should be taken.

Directions as to the observations, which may easily and profitably be made with regard to the present and past nature and extent of glacial action, the rate of movement of glaciers and the advance or retreat of their extremities, the snow-level, the extent and limit of forests and plants in mountain districts, and the relations of ranges to winds, rainfall, and climate, will be found in other sections.

General information on many subjects, both scientific and practical, connected with mountaineering, is given in a compact form by the late Mr. John Ball in his Introduction to 'The Alpine Guide' (new and revised edition by W. A. B. Coolidge, published, separately, in 1889 under the title of 'Hints to Alpine Travellers'), in the "Introductory Sections" to Murray's 'Switzerland' (Edition 1904), and the *Badminton* Volume on Mountaineering (1900). The last-mentioned book should be studied carefully by any traveller proposing to himself serious mountain exploration. He will find special chapters devoted to "Mountaineering beyond the Alps."

The Alpine Club has made arrangements to furnish information and advice to travellers who intend to visit mountainous countries. Application should be made to the Assistant Secretary, Alpine Club, 23 Savile Row.

Canoeing and Boating.—*By* J. Coles.

Canoeing.

Choice of a Canoe.—In making choice of a canoe the traveller must bear in mind that, in all probability, there will be rapids in the river, which will necessitate a portage being made, and that the canoe may have to be carried over rough ground for a considerable distance. For this reason, it is far better to take two canoes of moderate size than one large one, besides which, a small canoe is much more easily handled in bad water, and even should it become necessary to carry a large load, this can easily be done by lashing two small canoes together, at about one yard apart, and laying a platform across them, on which to place the stores, &c. This, however, should not be done in dangerous and rapid rivers. The following remarks do not, therefore, apply to large canoes, which, having nearly the stability of a boat, may be handled in the same manner.

Paddles.—It will generally be found that the native paddles will be best suited for the work. The double-bladed paddle, such as is used with canoes in this country, is quite useless on a rapid and dangerous river.

Sail.—The sail should be made of duck, or some such light material, fastened to a light yard at each end, and its hoist should be about twice its breadth; its size must be in proportion to the canoe, the hoist being about one-fourth of the canoe's length. The mast should be as light as possible, with a hole at the top for the halliards to pass through freely. The end should be stepped in a chock in the bottom of the canoe (when in use), and it should be lashed to one of the stays, or cross-pieces of the canoe. The sail should never be used unless the wind is steady and abaft the beam, and the halliards should be taken to the after part of the canoe in order to stay the mast, and secured in such a manner that it can be instantly let go, when the sail will at once fall, and undue pressure on the canoe relieved.

The Tow-line.—Too much attention cannot be paid to this important article. It should be light, but of the best material (such as the rope used by the Alpine Club), as its giving way at a critical moment in a rapid is sure to be attended with most serious results.

Loading the Canoe.—The packages should not exceed 50 lbs. in weight, as they may have to be carried long distances over portages, and care must be taken not to overload the canoe. Natives, who are all good swimmers, and have nothing to lose by a capsize, are very apt to put more into a canoe than is safe, so that it is a matter in which the traveller should use his own discretion.

In ascending a rapid river, keep close to one of its banks, and endeavour to take advantage of eddies. It will often happen that, owing to the strength of the stream, no headway can be made with the paddles, in which case recourse must be had to poling or tracking. In the event of the former, the poles should be straight and tough, and as long as can be conveniently carried in the canoe. Natives generally stand up to pole, but this the traveller should not attempt to do, or he will in all probability either fall overboard, or capsize the canoe, or both. In tracking, as great a length of line as possible should be used, as a sheer of the canoe in a rapid, with a short line, will often end in a capsize. Only two men should remain in the canoe, one in the bow with a pole, and the other in the stern with a paddle to steer; this man should also have his pole handy. The line should be made fast to one of the stays in the bow of the canoe, and *never to a towing mast*, as in a boat; as in passing round bad corners, or places where there are snags, and where it is necessary to give the canoe a wide sheer, the leverage of the mast, if the line were fastened to the top of it, would pull the canoe over. The man in the bow, however, should always have his knife handy to cut the tow-line, should necessity arise for his doing so. In tracking, when a river passes through sandy soil, the men on the line should keep at some little distance from the edge of the bank, as it is likely to give way under their weight, and precipitate them into the river. Several men lost their lives in Fraser River, in the early days of the gold discovery, by neglecting this precaution.

In crossing from one bank of a river to the other above a rapid, be careful to ascend the river for a considerable distance before attempting to do so; and then make the crew paddle as hard as they can, keeping the head of the canoe, if anything, rather down the stream, as in the case of a rapid river you would only lose ground by trying to fight against it.

In descending a river, the traveller should keep a look-out ahead for snags and places where the river is narrowed in between hills, as in such places there is nearly sure to be a rapid which may be so bad as to render navigation impossible. In all cases before descending an unknown rapid, he should land and inspect it throughout *its entire length* before attempting to run it in the canoe. When descending a rapid, care must be taken to keep steerage way on the canoe, as this will be needed to avoid rocks, or whirlpools. These latter are very serious dangers, as they generally do not remain fixed in one spot, but move about within a certain distance of a centre. There are, however, in most cases, short intervals when they break up, and that is the time to make a dash past them. To attempt this when they are in full swing could only end in the loss of the canoe and its occupants.

Boating.

When a traveller has to proceed for some distance overland before reaching a river or lake he purposes to navigate, he must of necessity provide himself with a boat constructed in such a manner as to be easily transported, either by being built in sections, that can be put together and taken to pieces at pleasure, or by taking one of the collapsible boats, such as Berthon's. If the former, he cannot do better than to have one built of Spanish cedar, on the same plan as that which was constructed for Sir H. M. Stanley, by Mr. James Messenger, of Teddington, with such modifications as may be necessary, when the means of transport, and the nature of his journey, have been duly considered. Collapsible boats, though very useful for ferrying across lakes or rivers, cannot, where a boat of other construction is available, be recommended for a continued exploration; they are, however, constructed of different sizes, and full particulars concerning them can be obtained from the Berthon Boat Co., 50 Holborn Viaduct, E.C.

If the exploration is to be commenced at the mouth of the river, a whale-boat will be found to be the best form of boat for the following reasons. Being steered by an oar, it is more easily handled in surf or a rapid; it is generally faster than boats of the same size of ordinary build; it will carry a good cargo, sail well off the wind, and is the best boat built for crossing the bars of rivers, or landing through a surf. Such a boat can generally be purchased at foreign ports, with oars and sail, and should be well overhauled before starting.

Boat-sailing cannot be taught by any book, and certainly not by a few short notes of this description. The traveller, therefore, who intends using a boat for exploration, should gather some experience before starting, which can be done at any fishing village on the coast. This will be the more necessary if he intends to use his boat on a lake, or for sailing along the coast, from the mouth of one river to another, and the following hints may, it is hoped, be useful to those who have had but small experience in boat-sailing.

When under sail, never, *under any circumstances*, allow the sheet to be made fast; a turn should be taken round a cleat, and it should be held by one of the crew ready to let go at any moment. Do not let the crew stand up, or sit on the gunwale. When about to round-to, remember that you cannot carry the same canvas on a wind that you can before it. If caught in a squall, put down the helm at once, ease the sheet, and if the squall is a bad one, lower the sail while it is still shaking. When approaching a danger, such as a rock, do not stand on if you are in doubt about weathering it, but go about in time, and have an oar ready to help the boat round if she appears likely to miss stays. Never carry too much sail, as there is considerable danger in doing so, and a boat will often sail faster with a reef taken in, than she will when unduly pressed. If necessary to take in a reef when sailing *on a wind*, do not luff, but check the sheet, lower the sail sufficiently to shift the tack, gather the sheet aft so that the men may take in the reef without leaning over the gunwale, shift the sheet, hoist the sail, while the sheet is slack, and do not haul the sheet aft until the men are again in their places.

Rowing.—This can only be acquired by practice, and though the traveller will seldom be called on to take an oar himself, circumstances may arise when he may have to do so, and we would, therefore, advise him to learn how to handle an oar before leaving England. Under ordinary circumstances, rowing on a river is sufficiently simple, and calls for no special instructions. The case, however, is very different when a river bar has to be crossed, or a landing made on a beach where a surf is breaking, and in either case it will be well to remember the following hints. On approaching the shore, a surf when seen from seaward never looks so bad as it really is. Where possible, a landing should not be attempted until opposite a village where the natives will be ready to assist the moment the boat touches the beach. When the surf is heavy, the boat should be backed in, pulling a few strokes to meet each heavy sea, and then backing in again until the shore is reached. The great thing to avoid is, letting the boat get broadside to the sea, as she will then capsize; a steer-oar should always be used, as a rudder is of little use in a surf, when backing in.

In crossing a bar, if there is a good, strong, fair wind, it will generally be best to cross under sail; but if the wind is light or variable, this should never be attempted. When rowing, the crew should be cautioned to keep their oars out of the water when the sea breaks round the boat, and to commence rowing again as quickly as possible afterwards. As even in the most experienced hands a boat will often be swamped on a bad bar, it will be well, before attempting to cross it, to prepare for a swim by removing all superfluous clothing, and see that everything that will float in the boat should be left free to float, while things that will sink, such as fire-arms, &c., should be securely fastened to the thwarts.

The remarks given on canoeing with regard to loading, to ascending and descending rapid rivers, are equally applicable to boating under similar circumstances, with the following exceptions. In towing, a short mast should be used to which the line is made fast; this is stepped in the same place as the mast, and should be stayed, so as to resist the strain of the tow-line. Paddles will often be found useful in reedy rivers where the oars get entangled. As a whale-boat empty will weigh about five hundredweight, more care must be taken at portages than in the case of a canoe, which can be lifted bodily over obstacles. The stems of small trees, or the oars should be laid down under the boat, and, where possible, sharp rocks must be avoided or moved out of the way. In a rapid, two men should be in the bow with poles ready to fend off from rocks, and the most experienced man of the crew should be in the stern with the steer-oar.

Although in the foregoing remarks special reference has been made to whale-boats, the hints given are equally applicable to boats of other construction, which should, however, for river work, crossing a bar, or landing through a surf, be fitted with a steer-oar in addition to the rudder. Awnings should be taken, but in rapid rivers, and when under sail, they cannot be used.

ORTHOGRAPHY OF GEOGRAPHICAL NAMES.

In 1878 the Council of the R.G.S., impressed with the necessity of endeavouring to reduce the confusion existing in British maps with regard to the spelling of geographical names, in consequence of the variety of systems of orthography used by travellers and others to represent the sound of native place-names in different parts of the world, formally adopted the general principle which had been long used by many, and the recognition of which had been steadily gaining ground, viz., that in writing geographical native names vowels should have their Italian significance, and consonants that which they have in the English language.

This broad principle required elucidation in its details, and a system based upon it was consequently drawn up with the intention of representing the principal syllabic sounds.

It will be evident to all who consider the subject, that to ensure a fairly correct pronunciation of geographical names by an English-speaking person, an arbitrary system of orthography is a necessity. It is hardly too much to say that in the English language every possible combination of letters has more than one possible pronunciation. A strange word or name, even in our own language, is frequently mispronounced. How much more with words of languages utterly unknown to the reader.

The same necessity does not arise in most Continental languages. In them a definite combination of letters indicates a definite sound, and each nation consequently has spelt foreign words in accordance with the orthographic rules of its own language.

It was therefore not anticipated that foreign nations would effect any change in the form of orthography used in their maps, and the needs of the English-speaking communities were alone considered.

The object aimed at was to provide a system which should be simple enough for any educated person to master with the minimum of trouble, and which at the same time would afford an approximation to the sound of a place-name such as a native might recognise. No attempt was made to represent the numberless delicate inflexions of sound and tone which belong to every language, often to different dialects of the same language, for it was felt not only that such a task would be impossible, but that an attempt to provide for such niceties would defeat the object.

The adoption by others of the system thus settled has been more general than the Council ventured to hope.

The charts and maps issued by the Admiralty and War Office have been, since 1885, compiled and extensively revised in accordance with it. The Foreign and Colonial Offices have accepted it, and the latter has communicated with the Colonies, requesting them to carry it out in respect to names of native origin.

Even more important, however, than these adhesions is the recent action of the Government of the United States of America, which, after an exhaustive inquiry, has adopted a system in close conformity with that of the R.G.S., and has directed that the spelling of all names in their vast territories should, in cases where the orthography is at present doubtful, be settled authoritatively by a Committee appointed for the purpose.

The two great English-speaking nations are thus working in harmony.

Contrary to expectation, but highly satisfactory, is the news that France and Germany have both formulated systems of orthography for foreign words, which in many details agree with the English system.

RULES.

The Rules referred to are as follows:—

1. No change is made in the orthography of foreign names in countries which use Roman letters: thus Spanish, Portuguese, Dutch, etc., names will be spelt as by the respective nations.

2. Neither is change made in the spelling of such names in languages which are not written in Roman character as have become by long usage familiar to English readers: thus Calcutta, Cutch, Celebes, Mecca, etc., will be retained in their present form.

3. The true sound of the word as locally pronounced will be taken as the basis of the spelling.

5. An approximation, however, to the sound is alone aimed at. A system which would attempt to represent the more delicate inflexions of sound and accent would be so complicated as only to defeat itself. Those who desire a more accurate pronunciation of the written name must learn it on the spot by a study of local accent and peculiarities.

5. *The broad features of the system* are:—

(*a*) That vowels are pronounced as in Italian and consonants as in English.

(*b*) Every letter is pronounced, and no redundant letters are introduced. When two vowels come together, each one is sounded, though the result, when spoken quickly, is sometimes scarcely to be distinguished from a single sound, as in *ai*, *au*, *ei*.

(*c*) Two accents only are used. (1) The acute, to denote the syllable on which stress is laid. The use of this accent is very important, as the sounds of many names are entirely altered by the misplacement of this "stress." (2) The sign ˘ with the vowel U, to indicate that the sound is open, as in *up*, not as in *pull;* as Tŭng, pronounced as in the English word *tongue*.

6. Indian names are accepted as spelt by the Survey of India.

7. In the case of native names in countries under the dominion of other European Powers in whose maps, charts, &c., the spelling is given according to the system adopted by that Power, such orthography should be as a rule disregarded, and the names spelt according to the British system, in order that the proper pronunciation may be approximately

known. Exceptions should be in cases where the spelling has become by custom fixed, and occasionally it may be desirable to give both forms.

8. Generic geographical terms, *e.g.*, those for Island, River, Mountain, etc., should be as a rule given in the native form. In the case of European countries, translation into English, where this has been the custom, should be retained, *e.g.*, Cape Ortegal, not Cabo Ortegal, River Seine, not Fleuve Seine.

9. For Chinese names, the Wade system of spelling is adopted.

N.B.—On any printed map or MS. document, an explanatory table, giving the English equivalents of the generic terms used, should of necessity be inserted.

The following amplification of these rules explains their application:—

Letters.	Pronunciation and Remarks.	Examples.
a	*ah*, *a* as in *father*	Java, Banána, Somáli, Bari.
e	*eh*, *a* as in *fate*, *e* in *benefit*	Tel-el-Kebír, Oléleh, Yezo, Medina, Levúka, Peru.
i	English *e*: *i* as in *ravine*; the sound of *ee* in *beet*. Thus, not *Feejee*, but	Fiji, Hindi.
o	*o* as in *mote*	Tokyo.
u	long *u* as in *flute*; the sound of *oo* in *boot*. *oo* or *ou* should never be employed for this sound. Thus, not *Zooloo*, but	Zulu, Sumatra.
	The shorter sound of the different vowels, when necessary to be indicated, can be expressed by doubling the consonant that follows. The sounds referred to are as follows:— The short *a*, as in *fatter*, as compared with the long *a*, as in *father* The short *e*, as in *better*, as compared with the long *e*, as in *fate* The short *i*, as in *sinner*, as compared with the long *i*, as in *ravine* The short *o*, as in *sobbing*, as compared with the long *o*, as in *sober* The short *u*, as in *rubber*, as compared with the long *u*, as in *rubric*	Yarra, Tanna, Mecca, Jidda, Bonny.*

* The *y* is retained as a terminal in this word under Rule 2 above. The word is given as a familiar example of the alteration in sound caused by the second consonant.

Letters.	Pronunciation and Remarks.	Examples.
	In the case of two different consonants following a short *u*, the accent *ŭ* may be used instead of doubling the consonant, as *Tŭng*, pronounced *tongue*.	
	Doubling of a vowel is only necessary where there is a distinct repetition of the single sound.	Nuulúa, Oosima.
ai	as in *aisle*, or English *i* as in *ice*	Shanghai.
au	*ow* as in *how*. Thus, not *Bowchee*, but	Bauchi.
ao	is slightly different from above	Macao.
aw	when followed by a consonant or at the end of a word, as in *law*	Dawna, Saginaw
ei	is the sound of the two Italian vowels, but is frequently slurred over, when it is scarcely to be distinguished from *ei* in the English *eight* or *ey* in the English *they*.	Beirút, Beilúl.
b	English *b*.	
c	is always soft, but is so nearly the sound of *s* that it should be seldom used. If *Celébes* were not already recognised it would be written *Selébes*.	Celébes.
ch	is always soft, as in *church*	Chingchin.
d	English *d*.	
f	English *f*. *ph* should not be used for the sound of *f*. Thus, not *Haiphong*, but	Haifong, Nafa.
g	is always hard. (Soft *g* is given by *j*) ..	Galápagos.
h	is always pronounced when inserted.	
hw	as in *what*; better rendered by *hw* than by *wh*, or *h* followed by a vowel, thus *Hwang ho*, not *Whang ho*, or *Hoang ho*.	Hwang ho, Ngan hwei.
j	English *j*. *Dj* should never be put for this sound.	Japan, Jinchuen.
k	English *k*. It should always be put for the hard *c*. Thus, not *Corea*, but	Korea.
kh	The Oriental guttural	Khan.
gh	is another guttural, as in the Turkish ..	Dagh, Ghazi.
l m n	As in English. When to represent the liquid sound of *l* in native names the French use the terminative *illes* (as in *Marseilles*), the *les* should be omitted. No attempt is made to represent the French nasal *n*.	

Letters.	Pronunciation and Remarks.	Examples.
ng	has two separate sounds, the one hard as in the English word *finger*, the other as in *singer*. As these two sounds are rarely employed in the same locality, no attempt is made to distinguish between them.	
p	As in English.	
ph	As in *loophole*	Chemulpho, Mokpho.
th	stands both for its sound in *thing*, and as in *this*. The former is most common.	Bethlehem.
q	should never be employed; *qu* (in *quiver*) is given as *kw*. When *qu* has the sound of *k* as in *quoit*, it should be given by *k*.	Kwangto.
r s sh t v w x	As in English. The *r* should be rolled or trilled.	Sawákin.
y	is always a consonant, as in *yard*, and therefore should never be used as a terminal, *i* or *e* being substituted as the sound may require. Thus, not *Mikindány*, *wady*, but not *Kwaly*, but	Kikúyu. Mikindáni, wadi. Kwale.
z	English *z*	Zulu.
zh	The French *j*, or as *s* in *treasure*	Muzhdaha.
	Accents should not generally be used, but where there is a very decided emphatic syllable or stress, which affects the sound of the word, it should be marked by an *acute* accent.	Tongatábu, Galápagos, Paláwan, Saráwak.

Russian Names.

The following table gives the equivalents of the Russian letters in the transliteration of Russian names:—

Printed Characters.	Cursive Characters.	Equivalents in R.G.S. System.	Remarks.	Printed Characters.	Cursive Characters.	Equivalents in R.G.S. System.	Remarks.
А а	*А а*	*a*		Т т	*Т т*	*t*	
Б б	*Б б*	*b*		У у	*У у*	*u*	
В в	*В в*	*v*		Ф ф	*Ф ф*	*f*	
Г г	*Г г*	*g(h)*	If *g*, always hard. If *h*, as in English.	Х х	*Х х*	*kh*	
Д д	*Д д*	*d*		Ц ц	*Ц ц*	*tz*	
Е е	*Е е*	*e*	*e* in benefit (*ye*, when initial).	Ч ч	*Ч ч*	*ch*	
Ж ж	*Ж ж*	*zh*	Sound of French *j*, or *z* in azure.	Ш ш	*Ш ш*	*sh*	
З з	*З з*	*z*		Щ щ	*Щ щ*	*shch*	*shch* in Parish church.
И и	*И и*	*i*		Ъ ъ	*Ъ ъ*	*mute*	Omit in transliteration.
І і	*І і*	*i*		Ы ы	*Ы ы*	*ui* / *i*	In middle of a word. / At end of a word.
К к	*К к*	*k*		Ь ь	*Ь ь*	*mute*	Omit in transliteration.
Л л	*Л л*	*l*		Ѣ ѣ	*Ѣ ѣ*	*ye*	
М м	*М м*	*m*		Э э	*Э э*	*e*	*a* in fate.
Н н	*Н н*	*n*		Ю ю	*Ю ю*	*yu*	
О о	*О о*	*o*	*o* in lock.	Я я	*Я я*	*ya*	
П п	*П п*	*p*		Ѳ ѳ	*Ѳ ѳ*	*f*	
Р р	*Р р*	*r*		Ѵ ѵ	*Ѵ ѵ*	*œ*	Seldom used.
С с	*С с*	*s*	Invariably sharp as a double *ss*.	Й й	*Й й*	*mute*	Omit in transliteration.

Note.—The termination ый and ій (adjective) should be transliterated *i*.

ON THE GIVING OF NAMES TO NEWLY-DISCOVERED PLACES.

The Council of the Royal Geographical Society would urge upon all travellers that in giving names to any new discoveries which they may make they should be guided by the following restrictions, which, until comparatively recent years, were commonly observed:—

1. That before putting forward any personal or fanciful name the traveller should do his best to ascertain that no local name exists, and where none is forthcoming should further consider whether one might not conveniently be derived from the vicinity, *e.g.*, from an adjacent stream, or pasture, or glacier, or from some characteristic of the natural object itself.

2. That no one should commemorate himself in this manner.

3. That any new nomenclature which a traveller may desire to suggest should be put forward tentatively and subject to the approval (1) of the Administration of the region or country, if there is one; (2) of the Official Cartographer of the country, if it possesses a Survey Department, or of the State to which the region may belong; or (3) of the Council of the Royal Geographical Scoiety.

INDEX.

See also page v, Vol. I.

[*To face last p. of Index.*

ADMISSION OF FELLOWS TO THE ROYAL GEOGRAPHICAL SOCIETY.

Candidates for admission into the Society must be proposed and seconded by Fellows, and it is necessary that the description and residence of such Candidates should be clearly stated on their Certificates.

It is provided by Chapter IV., § 1, of the Regulations, that,

> "Every Ordinary Fellow shall, on his election, be required to pay £5
> "as his admission fee, and £2 as his first annual subscription, or he may
> "compound, either at his entrance by one payment of £35, or at any
> "subsequent period on the following basis:—
>
> | | | |
> |---|---|---|
> | Fellows of 20 years' standing and over | .. .. | £12 10*s*. |
> | ,, 15 ,, ,, and under 20 | .. | £16 |
> | ,, 10 ,, ,, ,, 15 | .. | £20 |
>
> "And no Fellow shall be entitled to vote or to enjoy any other privilege
> "of the Society so long as he shall continue in arrear."

All subscriptions are payable in advance, on the 1st of January in each year.

The privileges of a Fellow include admission (with one Friend) to all ordinary Meetings of the Society, and the use of the Library and Map-room. Each Fellow is also entitled to Receive a copy of **The Geographical Journal**, which is forwarded, free of expense, to addresses in the United Kingdom or abroad.

Copies of the Regulations and Candidates' Certificates may be had on application at the Society's Office, 1, Savile Row, London, W.

INSTRUCTIONS FOR INTENDING TRAVELLERS.

Instructions for Intending Travellers, under the authority of the Council of the Royal Geographical Society.—Arrangements have been made for the instruction of intending Travellers in the following subjects:—

1. Surveying and Mapping, including the fixing of positions by Astronomical Observations. By Mr. E. A. Reeves, F.R.A.S., Map Curator of the Society.

   Those pupils who have gone through Mr. Reeves's Course, and received a Certificate from him, can obtain instruction in the following additional subjects:—

2. Geology, including practical training in the field. Under the superintendence of the Director-General of the Geological Survey.
3. Botany. Under the superintendence of the Director, Royal Gardens, Kew.
4. Zoology. By Dr. R. Bowdler Sharpe, Natural History Museum, South Kensington.
5. Anthropological Measurements. Under the superintendence of the Anthropological Institute.
6. Photography. By Mr. John Thomson, Author of "Photographic Illustrations of China and its People," and other works.
7. Meteorology. By Dr. H. R. Mill.
8. Outfit and Health. By Charles Forbes Harford, M.A., M.D., Secretary of the Travellers' Health Bureau.

The lessons are given on days and at hours arranged between the Instructor and the pupil.

Tickets for the lessons, at 2*s.* 6*d.* for each hour's lesson, must be previously procured at the Offices of the Society.

LONDON: PRINTED BY WILLIAM CLOWES AND SONS, LIMITED,
DUKE STREET, STAMFORD STREET, S.E., AND GREAT WINDMILL STREET, W.

Breinigsville, PA USA
28 December 2010

252305BV00002B/1/P